Wisconsin State Parks

Sandstone outcropping in
Rocky Arbor State Park

WISCONSIN STATE PARKS

Extraordinary Stories of Geology and Natural History

SCOTT SPOOLMAN

WISCONSIN HISTORICAL SOCIETY PRESS

Published by the Wisconsin Historical Society Press
Publishers since 1855

The Wisconsin Historical Society helps people connect to the past by collecting, preserving, and sharing stories. Founded in 1846, the Society is one of the nation's finest historical institutions.
Join the Wisconsin Historical Society: wisconsinhistory.org/membership

Photographs are by Scott Spoolman unless otherwise noted.
The front cover photo shows the view from the East Bluff of Devil's Lake State Park, with Lake Wisconsin visible in the distance. Used with permission of Derrick Mayoleth, www.devilslakewisconsin.com.

Printed in Canada
Designed by Percolator Graphic Design
27 26 25 24 6 7 8 9

Library of Congress Cataloging-in-Publication Data

Names: Spoolman, Scott, author.
Title: Wisconsin State Parks : extraordinary stories of geology and natural history / Scott Spoolman.
Description: Madison, WI : Wisconsin Historical Society Press, [2018] | Includes bibliographical references and index. |
Identifiers: LCCN 2017041545 (print) | LCCN 2017056929 (e-book) | ISBN 9780870208492 (paperback) | ISBN 9780870208508 (e-book)
Subjects: LCSH: Parks—Wisconsin—History. | BISAC: SCIENCE / Earth Sciences / Geology. | TRAVEL / Hikes & Walks. | HISTORY / United States / State & Local / Midwest (IA, IL, IN, KS, MI, MN, MO, ND, NE, OH, SD, WI).
Classification: LCC SB482.W6 (e-book) | LCC SB482.W6 S66 2018 (print) | DDC 363.6/809775—dc23
LC record available at https://lccn.loc.gov/2017041545

♾ The paper used in this publication meets the minimum requirements of the American National Standard for Information Sciences—Permanence of Paper for Printed Library Materials, ANSI Z39.48-1992.

This book is dedicated to my parents, Art and Betty Spoolman,
who cherished hunting for blueberries, wildflowers, and waterfalls,
and who taught us to love Wisconsin's natural beauty as much as they did.

Contents

Fractured bedrock near the Niagara Escarpment in High Cliff State Park

Preface

If you could view Wisconsin from outer space on a sunny, clear day, it might look something like a relief map, with areas of rough terrain in brown and gray, smoother green expanses, rambling river valleys and their tributaries, and blue lakes scattered across the state and bordering its north and east sides. These varied landscapes can be traced to the complex geologic story of the state, which in fact is not just one story but several. Wisconsin's state parks make perfect entry points to these stories, as places where visitors can see the variety of land formations and other evidence of billions of years of geologic and natural history.

This book is organized geographically, starting in the northwest corner of the state and finishing in Door County. This order of presentation conforms with the state's geologic story to some extent. The northwest corner of Wisconsin is underlain by one-billion-year-old volcanic rock that is exposed in many places. On much of the eastern side of the state, Wisconsin's youngest bedrock lies at the surface, and that side of the state was greatly affected by the glaciers, which were the most recent geologic events. I followed the general order of the chapters in my travels to state parks and in the research and writing process. This gave me a good grasp of the big picture of geology and natural history in Wisconsin, and I hope it aided my effort to share my fascination with my home state's ancient history.

It is not necessary to follow this order in reading the chapters. However, I strongly recommend that readers start with the first chapter's overview of Wisconsin's geologic history before reading other chapters. This introductory chapter includes important basic material on geology and natural history, along with the timeline of major geological events and processes that will help readers understand the region-specific information. Each chapter begins with an introduction that goes deeper into the geologic processes that affected the geographic area covered. Each park story then completes this zooming-in process, sharing even more details about what happened there. Finally, each park story is followed by one or more trail guides that help the reader to find evidence on the ground of what has been described.

Because this book is not intended to be a guidebook in the vein of most published hiking guides, the trail guides are not exhaustive or detailed in terms of trail difficulty, exact distances, associated amenities, and other such details. With their focus on the geology and natural history of each park area, these trail guides are intended to serve as companions to other sources such as state park maps, drawing hikers' attention to features in the parks that help to tell each story of the ancient and more recent past.

Wisconsin has a rich diversity of more than 100 state parks, state forests, and state recreation areas, each of which has its own story. Not all of these stories are about geology and natural history. Some sites showcase more recent history while others are focused on recreational opportunities such as boating, swimming, and bird-watching. Choosing which parks to feature in this book was not easy. I selected them with the hope of providing the best sampling of the major geological processes and events that affected the state and the resulting natural and early human historical developments for each site. With that in mind, I chose 26 state parks and two state forests.

Several of the state parks include or overlap with state natural areas (SNAs). The SNA Program operates within the Bureau of Endangered Resources of the Wisconsin Department of Natural Resources. It was created in 1951 with the goal of preserving native plant and animal communities that were rapidly disappearing due to settlement and development of the state's wild areas. Several of these communities, or in some cases certain species living there, are rare or endangered. Some SNAs also protect fragile geological formations, and others protect archaeological sites. In 2017, there were 683 SNAs across the state. They serve as areas of study for scientists and students and receive more protection than other public lands get. Park visitors are asked to stay on marked trails, especially when crossing SNAs.

Tying all of the state park stories together are the connections between geology and natural history and human history in Wisconsin. What is under the Earth's surface often shapes what goes on above. For example, some sandy areas of the state—the legacy of ancient seas that deposited deep beds of sandstone—support certain types of forest ecosystems that in turn host certain animal communities. The ways in which early humans used resources from these ecosystems shaped their lifeways and communities over time. The same is true of how these resources shape communities and economies today. We can see these connections time and again in Wisconsin's state parks and their surrounding areas.

1.1 Relief map showing the varied landscapes of present-day Wisconsin, shaped by billions of years of geologic processes. WISCONSIN GEOLOGICAL AND NATURAL HISTORY SURVEY

1

The Big Picture

An Overview of Wisconsin's Geological and Natural History

To understand the geological history of any place, it's necessary to shift out of the daily time frame that is measured in minutes, seconds, and even smaller and smaller fractions of a second. The overall history of planet Earth reaches back 4.6 billion years. That's 4,600 millions of years, or 46 million centuries. Making the shift from a contemporary to a geologic time frame means thinking of a million years as a brief period and of centuries as flashes in time.

Geologists think that mountains have risen and disappeared in what is now Wisconsin. If that seems hard to believe, try dividing the total geologic time frame into segments of 50 million years. One of those segments is 500,000 centuries. Considering all that happens in a single century, it's easier to imagine a great deal more could occur in half a million centuries.

In fact, geologists estimate that it took between 50 million and 100 million years for a great mountain range to be formed across parts of Minnesota, Wisconsin, and Michigan. Nearly 2 billion years ago, two continents collided and heaved that land up, but it was an ultra-slow-motion process. We do not know how high they rose, but assume for the purposes of illustration that over a period of 75 million years some of those mountains reached a height of 15,000 feet. That means they would have grown at an average rate of less than three one-thousandths of an inch—about the thickness of an average fingernail—per year, which certainly seems believable.

Similarly, if wind and rain could wear away an average of just two one-thousandths of an inch of such a mountain every year, it would be gone within 100 million years once the forces of erosion took over. Indeed, Wisconsin's mountains were gone within that period of time, according to the evidence. Over billions of years, many such drastic changes are possible.

Geologists are trained to shift into the geologic time frame, and they have done exhaustive work searching through the geologic record. Based on hundreds of years of such research, they have woven a geological history of the land area that is now Wisconsin (Figure 1.1). What follows is a brief summary of that story, intended to be a framework within which the stories of Wisconsin's state parks can be told. Because geologists tend to rely on the million-year time unit, long time periods are usually expressed in millions of years throughout this book. For example, 4 billion years ago is 4,000 million years ago (MYA) (Figure 1.2).

Precambrian Era (4,600 – 545 MYA)

Precambrian is the name given to the vast stretch of time before evidence of life began to accumulate in fossils—the remains, traces, or imprints of ancient organisms preserved in Earth's crust.[1] The Precambrian era occupied more than 4,000 million of Earth's 4,600-million-year history. Early in Precambrian time, the planet's crust was segmented into several huge, irregularly shaped tectonic plates. Since then, these plates have been shifting around atop Earth's plastic mantle layer, driven by the intense and turbulent heat of the core and mantle. This continuing process is called plate tectonics.

The border zone between any two plates is called a fault. Faults vary greatly in their nature. In some cases, plates are colliding while in others, they are spreading apart. Still others are sliding along next to each other in opposite directions. Faults are where the action is; volcanoes, earthquakes, and other complex geologic processes take place there. Through such processes, ancient continents formed on some of the Earth's plates. Tectonic forces jammed some of the continents together, split some of them apart, and reorganized these early land masses into continents whose sizes, shapes, and locations have varied over time.

3,500 MYA

The precursors to Wisconsin's oldest rocks were formed by the movement of magma, or hot, fluid rock, into fissures and chambers beneath the Earth's

AGE	PERIOD		YEARS AGO
Cenozoic	Quaternary	Holocene epoch	
			12,000
		Pleistocene epoch	
			2.6 mya
	Tertiary		
			65 mya
Mesozoic	Cretaceous		
			145 mya
	Jurassic		
			208 mya
	Triassic		
			248 mya
Paleozoic	Permian		
			286 mya
	Pennsylvanian		
			320 mya
	Mississippian		
			360 mya
	Devonian		
			417 mya
	Silurian		
			444 mya
	Ordovician		
			495 mya
	Cambrian		
			545 mya
Precambrian	Proterozoic eon		
			2,500 mya
	Archean eon		

1.2 The geologic time table. ADAPTED FROM DOTT AND ATTIG, *ROADSIDE GEOLOGY OF WISCONSIN*

surface and volcanic eruptions of magma onto its surface. The magma cooled and solidified, becoming igneous rock such as granite (formed underground) and basalt (formed on the surface). Geologic forces fractured and melted some of this rock miles below the surface where it was deformed and reconfigured by unimaginable heat and pressure to become metamorphic rock, part of the foundation of the continent.

2,900 MYA

Oxygen began to build in the atmosphere as Earth's first photosynthesizing organisms, cyanobacteria, grew on ocean surfaces. The land was barren, and with no vegetation to block wind or anchor soil, it was subject to massive erosion by wind and water.

2,500 MYA

One of the ancient continents of this time was the Superior continent, the southern shore of which included what is now far northern Wisconsin and Michigan's Upper Peninsula. It is also known as the Canadian Shield—the core of ancient North America. To the south across a sea lay a smaller land mass called the Marshfield continent. At this time, the whole region was south of the equator. Geologic evidence shows that the area was as flat and arid as the Persian Gulf area is today, still void of vegetation and other life-forms.[2]

Seas had been rising and falling for millions of years in this region, alternately covering the land, then receding. During underwater periods, the sea floor accumulated layers of sand and other sediments eroding from the dry highlands. Over millions of years, the increasing weight of these layers, along with certain chemical reactions, converted these sediments to sandstone, a form of sedimentary rock—the third major type of rock after igneous and metamorphic rock.

1,900–1,785 MYA

During this span of time, tectonic forces caused the plates underlying the Superior and Marshfield continents to collide. In this extremely slow-motion collision, the edge of the northern plate was subducted, or driven beneath the southern plate into the Earth's intensely hot mantle. This caused volcanic activity beneath the sea that built a row of volcanic islands, not unlike Alaska's Aleutian Islands, off the northern coast of the Marshfield continent.

As the plates continued to collide, the volcanic islands, pushed by the Marshfield continental plate, rammed the Superior continent about 1,860 million years ago. About 30 million years later, the southern edge of the volcano chain, then the south coast of the Superior continent, collided with the land mass of the Marshfield continent. These colossal events generated more heat and volcanic activity and more churning of the rocks. They also heaved up the land to create the Penokean Mountain range stretching across parts of present-day Minnesota, northern Wisconsin and Michigan, and southern Ontario. It was long, narrow, and lofty, on the order of the present-day High Sierra mountain range in California, but hot, dry, and barren.

1,760 MYA

Southern Wisconsin—then a rolling plain of ancient granite and greenstone—saw violent volcanic activity, including explosive volcanoes and underground flows of magma that created massive bodies of granite. Volcanic eruptions covered much of southern Wisconsin in ash and magma that became the granite-like rock rhyolite.

Meanwhile, for about 25 million years, the Penokean Mountains were being eroded by intense winds and precipitation. The sand and clay scoured off by this erosion were carried by streams flowing primarily southeast to an advancing sea, where they were deposited on the bedrock. These sediments accumulated in layers and eventually were forged into sandstone and shale, several thousand feet thick in some areas.

1,700–1,500 MYA

Near the southern shore of the continent, which was then somewhere in northern Illinois, a new, smaller mountain range appeared. It probably resulted from another continental collision, which compressed and folded the deep layers of sandstone.[3] Over about 70 million years, the heat and pressure from this folding process metamorphosed much of the sandstone, forming the hard rock quartzite. The remnants of the quartzite mountain range include the present-day Baraboo Hills.

Around 1,500 million years ago, another period of underground movement of magma produced a massive body of granite under much of east-central Wisconsin. Centered under the Wolf River, it is called the Wolf River batholith.

1,109 MYA

The Penokean Range had been almost completely leveled by 1,285 million years ago, and the region was mostly a gently rolling plain, still south of the equator and therefore tropical. Then, about 1,109 million years ago, in the area underlying present-day Lake Superior, a plume of magma burst toward the surface from somewhere deep in the mantle and began a major event known as the Midcontinent Rift. The magma plume fractured the surface in an arcing pattern. The area that is now Lake Superior was at the top of the arc, and the two legs stretched away to the southwest into Oklahoma and southeast into Alabama. The land areas on either side of this rift were propelled away from each other in what could have been the birth of an ocean much like the Atlantic Ocean, which was formed to the east much later by a similar process.

Within a few million years, something stopped the rifting process, possibly another continental collision to the southeast.[4] Instead of splitting in two, the continent remained whole, but within and around the rift, the action continued. From the rift, the lava kept erupting and spreading across the land, covering an area extending perhaps 100 miles on either side of the central rift zone with volcanic basalt.[5] This event played a major role in the formation of northern Wisconsin's landscape. Geologists refer to this rifting period as the Keweenawan (kee-wee-NAH-wan) episode, or Keweenawan time—named for Michigan's Keweenaw Peninsula, which extends into the Lake Superior basin.

Roughly during this same period, what is now the north-central part of Wisconsin, located between the two extensions of the rift area, was gently lifted into a dome-shaped land area, the Wisconsin Dome. It may have formed as the land on either side of it gradually subsided or sank. It is now commonly known as the state's Northern Highlands—represented in Figure 1.1 as the roughly circular brown and tan area in north-central Wisconsin.

1,000 MYA

The rifting had stopped and the separated plates were being pushed back toward one another, and this compressed the land mass within the Lake Superior basin. Due to this compression, areas of the formerly subsiding basalt and sediment layers were fractured into blocks that were heaved up. In some of these fault blocks, you can see layers of ancient rock next to layers of younger rock. In some areas, the older (formerly deeper) rock is actually lifted above the younger layers.

Probably due to a combination of the collision with the continent to the southeast and tectonic forces within the Earth, some areas of the continent warped upward and others downward, creating an arch-and-basin pattern across central North America. One of the arches was the Wisconsin Arch, part of the area between the two legs of the Midcontinent Rift. It ran south-southwest from the Wisconsin Dome to just west of what is now Madison (in Figure 1.1, the tan tail-like area stretching south from the dome). The basin to the east, called the Michigan Basin, contains Lake Michigan, Michigan's lower peninsula, and the main body of Lake Huron.

The violence of volcanic and tectonic activity died down during the last part of the Precambrian era, at which time erosion by wind and rain was the dominant force. The Wisconsin region was still tropical, being slightly south of the equator. Wisconsin was part of a vast, heat-blasted expanse of rock and sand, not unlike today's large deserts. With no vegetation, erosion was rapid and continued for hundreds of millions of years. Consequently, no geological record remains of what happened at this time in the region.

The Paleozoic Era (545–248 MYA)

545–495 MYA: The Cambrian Period

This first period of the next era of Earth's history began around 545 million years ago. In the oceans, plant and animal life began to evolve quickly. The land remained barren, possibly except for lichens clinging to rocks in moist places. However, during the last 400 million years of Precambrian time, primitive ocean life had generated enough oxygen to make the atmosphere suitable for the earliest land animal species, which appeared about 530 million years ago.

What is now Wisconsin was located just below the equator with its north-south axis turned 90 degrees and oriented east-west (Figure 1.3). (For simplicity, I use today's compass points for all directional references in this book.) With the coming of the Cambrian, seas began to advance across North America from the east and west. Eventually, all but perhaps the far northwest corner of the state was under water in a shallow tropical sea—probably about 100 to 300 feet deep on average—often referred to as the Cambrian sea.[6]

Broad, braided rivers flowing to this sea deposited sand and other materials on the shore. Where the rivers met the sea, they lost velocity and dropped

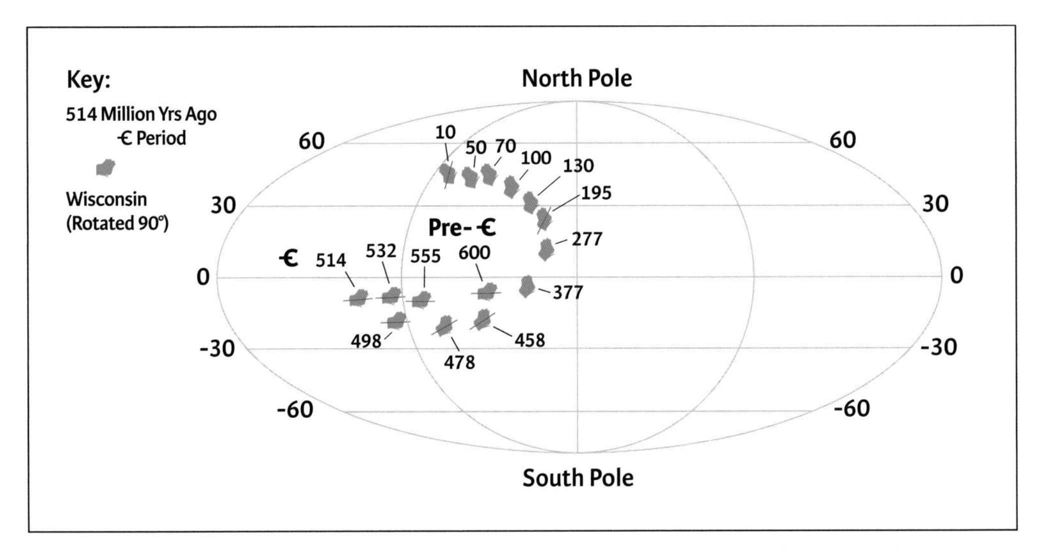

1.3 During the past 600 million years, the land area that is now Wisconsin moved from below the equator to its present location and orientation. Numbers represent millions of years ago. MICKELSON ET AL., *GEOLOGY OF THE ICE AGE NATIONAL SCENIC TRAIL*. REPRINTED BY PERMISSION OF THE UNIVERSITY OF WISCONSIN PRESS

heavier materials right there. The diminishing outflowing currents were able to carry sand grains a little farther. Beyond that limit, only the tiny particles that make up silt and clay could be carried much farther and were deposited as mud on top of the sand and gravel that had been laid down earlier.

On this ancient coast, storms brought waves that crashed against steeper shorelines, tearing chunks of rock from the land and pulling them into the sea. At these cliff sides, the turbulent water rolled and tumbled these rocks for thousands of years, until they became rounded and smoothed and sand filled in the spaces between them.

With the advance of the Cambrian sea, the shorelines inched across the land over hundreds to millions of years. Like infinitely patient plasterers, the streams flowing to the sea and the waves and currents flowing along the shorelines deposited layers of silt, sand, and gravel, grain by grain, troweling these sediments over the coastal land. When these deposited layers accumulated deeply enough, their collective weight along with certain chemical processes forged layers of sedimentary rock—sandstone from the sand layers, conglomerate from the layers of mixed gravel and sand, and shale from layers of mud.

Scientists think the shallow sea retreated and advanced at least twice during the Cambrian period. After the last Cambrian sea departed, most of Wisconsin was covered with sandstone and conglomerate, interspersed with layers of shale in some places. This old seabed was then subject to a few million more years of wind and water erosion that erased most of the geologic record of the period.

495–444 MYA: The Ordovician Period

After the Cambrian, seas advanced again to Wisconsin and beyond. They ushered in a more diverse collection of ocean life-forms, including brachiopods, corals, and clams, whose fossils can be found in parts of Wisconsin. Other organisms began building reefs—early versions of the stunning coral reefs found today in the tropical zone. The wastes and remains of ancient plants and animals deposited on sea floors contributed to the formation of two closely related types of sedimentary rock, limestone and dolostone, the latter commonly referred to as dolomite. Whereas the Cambrian sea sediments primarily formed sandstone, conglomerate, and shale, Ordovician sediments are more associated with dolomite.

During this 51-million-year period, the Ordovician sea retreated and advanced again at least once and perhaps four times. Each re-advance of the sea brought new deposits of limestone that would become dolomite. The result is a rock record that contains mostly dolomite with minor layers of sandstone and shale interspersed.

Another major development throughout Cambrian and Ordovician time was the continued rise of the Wisconsin Dome. This gentle upheaval—which geologists call upwarping—also expanded into south-central Wisconsin, lengthening the Wisconsin Arch.

444–417 MYA: The Silurian Period

During the next 27 million years, a distinctive set of life-forms evolved in the oceans. North America was still tropical, straddling the equator. The first major development of coral reefs took place in the clear, warm waters of the seas, especially in the Michigan Basin off the southeastern shore of present-day Wisconsin. The period is referred to as the Age of Corals.

For most of Silurian time, Wisconsin was probably under the sea, possibly except for a part of the northwest corner. Because of the rich diversity of sea life, a thick layer of limestone that was converted to dolomite was deposited over most of the state during Silurian time. When the seas withdrew, erosion once again became the dominant land-shaping force for millions of years, removing most of the hard Silurian dolomite and underlying Ordovician layers from much of Wisconsin. However, where dolomite resisted erosion and remains today—mostly in eastern Wisconsin—it played a key role in the geologic story.

417–360 MYA: The Devonian Period

During this last period in which tropical seas covered Wisconsin, sea life continued to evolve. Fishes appeared and evolved so quickly that this period of 57 million years is called the Age of Fishes. In rocks from the earliest part of the period, evidence of the first land plants has been found. By the end of the Devonian, primitive forests covered some land areas with 30-foot tree ferns and various forms of leafless plants. Air-breathing amphibians had emerged from the seas, and there began a rapid evolution of animal life, including amphibians, reptiles, and insects.

Some geologists hypothesize that Wisconsin was covered by seas intermittently during the Ordovician, Silurian, and Devonian periods and possibly for another 100 million years beyond, after which Paleozoic seas withdrew from the continent's interior for the last time.[7] Thus, Wisconsin has been high and dry for at least 260 million years. However, most of that later period, as well as the rest of the time until the beginning of the Ice Age, remains a geological mystery.

The Mesozoic Era
(248 – 65 MYA)

For the period between roughly 400 million and about 2 million years ago (which includes the latter half of the Paleozoic era), sea levels dropped around the world, and a diverse array of dinosaurs evolved and ruled the land. But virtually no rock record exists of what happened in the Wisconsin, thanks to the effects of erosion. Just as marauders who burned ancient libraries erased much of human history, wind, water, and ice have wiped clean much of the rock and fossil record after the Devonian period. For that reason, scientists can only speculate about what kinds of dinosaurs and other animals, as well as plants of the time, lived within the area that is now Wisconsin.

We do know that during this long period, Wisconsin was on the move, riding on its slowly rotating, drifting tectonic plate.[8] About 600 million years ago, what is now the north-south axis of the state was oriented east and west and the state's region lay south of the equator. About 350 million years ago, that axis had tilted to a northeast-southwest orientation, and Wisconsin had crossed the equator, heading north (Figure 1.3).

The Cenozoic Era (65 MYA – PRESENT)

65–2.6 MYA: The Tertiary Period

During the first period of the Cenozoic era, North America enjoyed a moist climate, milder than it is now, according to some geologists. Grasses, trees, and mammals were well established, and the land hosted a rich diversity of prairies, forests, and wetlands. The center of the continent during that time period has been compared to the grasslands of present-day Africa—teeming with life, including mammoths, giant ground sloths, camels, beavers, porcupines, brown bear, saber-toothed cats, and the ancestors of today's rhinoceroses, horses, dogs, and cats.[9] There is no evidence of the ancestors of humans living in the Wisconsin area at the time. Toward the end of Tertiary time, the climate began to cool as part of a global cycle.

2.6 MYA–Present: The Quaternary Period

The Quaternary period has been divided into two epochs—the Pleistocene (2.6 million years ago to about 12,000 years ago), and the Holocene (12,000 years ago to present). A little more than 2 million years ago, the mild climate changed. The northern land areas of the northern hemisphere cooled to the point where winter snow piled up enough to remain through the warm months of the year, and each year, more snow remained. As this supply of snow grew larger and deeper, the weight of overlying snow layers compacted those beneath them and formed ice layers that accumulated to hundreds of feet thick in certain areas. When these ice masses got to be thicker than about 100 feet, gravity began pulling them down slopes while the force of expanding ice helped them to spread over larger and larger areas as vast ice sheets, or glaciers. Thus began the coldest part of the Pleistocene epoch, also called the Ice Age, during which about 30 percent of the planet's land area was covered several times with ice (three times the amount covered today).

Thousands of years after the first advance of this ice from the north, the Earth's climate warmed and the glaciers melted and retreated to where they had first formed in the north. Then the climate cooled and the cycle began again. Geologists estimate that this glaciation cycle has occurred 10 to 15 times at roughly regular intervals.[10]

North America's largest glacier, the Laurentide Ice Sheet, originated in Canada around Hudson Bay, straight north of Wisconsin, and advanced south time and again. Year after year during each advance, the winter added another layer of ice to the glacier, giving it more mass and thus more power to expand south or in whatever direction the lay of the land allowed. As it grew, it inched across land, plowing up boulders, gravel, and sand and moving it all forward.

The most recent advance of this ice sheet occurred between roughly 85,000 and 11,000 years ago. At its peak, it extended northward to the Arctic Ocean and from the west coast to the east coast of Canada. It reached into the northern tier of the states and dipped as far south as Illinois and Missouri. It was named the Wisconsin glaciation because Wisconsin contains some of the region's best representations of that glacier's effects on the land.

This last advance of the glacier reached Wisconsin about 30,000 years ago, and 6,000 to 7,000 years later, it had covered all of the northern and eastern parts of the state. The ice flowed in separate lobes, or enormous tongues of ice. In Wisconsin the major lobes were named for the lowlands through which they flowed (Figure 1.4).

One of the features that makes Wisconsin geologically unique is its Driftless Area, occupying most of the southwest corner of the state. Years ago, geologists used the word *drift* as a collective term for the rocks and soil carried by glaciers and dropped on the land as they melted back. Wisconsin's Driftless Area has no evidence of drift, and geologists conclude that the area was never ice-covered. Scientists from around the world have studied the Driftless Area as a showcase of unglaciated landscape.

As the cold climate encroached upon the state from the north, forests and grasslands gradually migrated southward, dying away in the north and expanding to the south. The animals that lived in these environs went with them. When the ice advanced into the state, tundra conditions developed, and permafrost reached hundreds of feet into the ground. Warmer weather still occurred during the brief summers, and a few feet of ground thawed. This tundra supported only lichens, grasses, sedges, and a few flowering plants, but no trees.

The glaciers excavated the areas they covered, plowing away softer rock and leaving basins, piling up great masses of rock and soil to build ridges and hills, and carving the land to create stunning escarpments and promontories. The

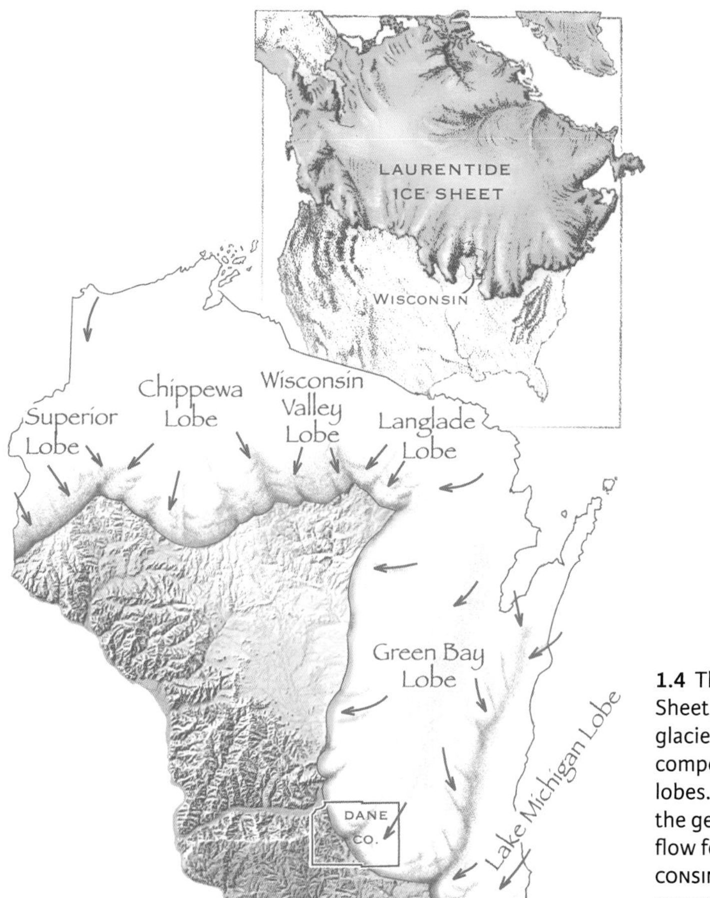

1.4 The Laurentide Ice Sheet, the most recent glacier in Wisconsin, was composed of six major lobes. Arrows indicate the general direction of flow for each lobe. WISCONSIN GEOLOGICAL AND NATURAL HISTORY SURVEY

Great Lakes were created largely by the Wisconsin glaciation. The basin that was gouged out to make room for Lake Superior reaches depths of more than 1,300 feet.

Even as they melted back when the climate warmed after each glaciation, the ice sheets affected the land dramatically. Meltwater flowed in the ice-free river valleys carrying untold tons of sand and gravel and building broad floodplains. The retreating ice temporarily dammed large volumes of meltwater, and when the dams broke, massive torrents of ice water gushed through the drainage routes, tearing away rock and carving exquisite landforms. When the ice was gone, it had dropped vast fields of clay, sand, gravel, and boulders known as till. Out of this material, the glaciers also built ridges and hills—called moraines, drumlins, kames, and eskers—as well as other landforms now on display across much of Wisconsin.

The climate began warming about eighteen thousand years ago. As the ice retreated to the north, frigid conditions lingered at its margin and the tundra climate returned. The temperature difference between the newly ice-free, warmer land and the still-frigid ice mass created strong winds. In the years or decades before any vegetation grew, these winds dried the land and roiled clouds of dust and sand, darkening the sky for weeks at a time and building vast sand dunes in some places.

The ice finally left northernmost Wisconsin about 9,500 years ago. Winds died down and plant communities slowly took hold. Grasses and other species whose seeds are carried by winds were the first to grow in these new ecosystems. Tundra slowly gave way to scattered open forests of spruce, fir, and jack pine. Soils developed, and after thousands of years came the species that depend on richer soils—white pine, maple, and eastern hemlock—forming well-established forest systems in some areas.[11] Prairies and savanna communities spread and flourished in other areas of the state.

The postglacial tundra hosted musk oxen, mastodons, wooly mammoths (Figure 1.5), elk, caribou, white-tailed deer, wolves, giant beavers (larger than most black bears), and other large mammals, or megafauna. Feeding on these warm-blooded species were hordes of mosquitoes. The megafauna moved north as the ice retreated, and later many such species disappeared in one of the Earth's largest extinction bursts. The forests and grasslands that replaced the tundra after thousands of years were home to small mammals including chipmunks, squirrels, mice, and foxes. Early residents of wetlands and lakes included loons, red-winged blackbirds, wood ducks, and muskrats, living among the cattails, sedges, and other wetland plants.[12] By about 5,000 years ago, Wisconsin looked much as it did before Europeans began settling and dramatically changing the land for farming and other purposes in the 1700s.

Humans had occupied the land long before the arrival of Europeans, of course. With the retreat of the glacier, these early peoples had entered North America via the Bering Land Bridge, which formed when much of the planet's water was tied up in ice and sea levels were low. Anthropologists call these people Paleo-Indians. These hunter-gatherers are thought to have entered Wisconsin about 12,000 years ago, following mammoths, mastodons, bison, giant ground sloths, and musk oxen as they roamed eastward and northward, crossing the continent along the margin of the glacier. As postglacial sea levels rose, the land bridge they had arrived on was flooded and now lies under the Bering Strait separating Alaska from Siberia.

The next group recognized by anthropologists was the Early Archaic people, who lived in the area between 10,500 and 8,000 years ago. With the changing landscape and the growth of forests, tribes were able to settle into smaller areas and become less nomadic. They continued the Paleo-Indian hunting tradition, but as the megafauna moved north and eventually disappeared, hunters relied more on small mammals, and people turned more to fishing and gathering wild plants. As their populations grew, these people settled into particular regions, adapting their ways of life to the varying landscapes and environments.

The Middle Archaic time, lasting from 8,000 to 5,000 years ago, saw further change in the direction of settling and working the land. As the climate continued to warm and forests grew deeper and more diverse, hunters pursued deer and other game. The people developed their plant-gathering tradition and included the harvesting of nuts. They also invented tools such as axes and celts (primitive chisels) for working with wood. Based on the tools found, anthropologists have speculated that these people may have built dugout canoes and made

1.5 The wooly mammoth roamed Wisconsin during and after the last glacial age. RACHEL KLEES-ANDERSEN

wooden bowls and other implements during this time, although such objects would have long since disintegrated into the soil.

Late Archaic peoples lived between 5,000 and 3,000 years ago. They developed new hunting tools, including spear throwers called atlatls. They used copper more widely for making tools and other goods. Hunting, fishing, and plant gathering all became more sophisticated and efficient. Planning and organization became more important in communities. These people also created and expanded trade networks among regions.

The next period, the Early Woodland, running from 3,000 to 2,300 years ago, saw a continuing shift toward more reliance on plants as a food source, in addition to hunting. People also developed pottery, burial mounds, and cultivation of crops. Agriculture probably began with the intentional tending of wild plants, from which the caretakers learned how to gather and plant seeds and harvest the fruits or vegetables. Early crops included beans, corn, and squash. Because of agriculture, populations became more concentrated, and leaders began to gain more status and power. Human burials became more ceremonial, and the use of mounds to hold human remains may have started during this period.

The Middle Woodland period—2,300 to 1,500 years ago—was characterized by the building of villages along rivers in the south and on lakeshores in the north. Village-centered life is thought to have enhanced hunting, fishing, plant gathering, and gardening. Fishing became more sophisticated with the use of gill nets, harpoons, and fishing hooks and lines. Also during this time, a people from south of Wisconsin called the Hopewell culture traveled north. They brought shells from the Gulf Coast, obsidian from the Yellowstone National Park area, items made from pearls and silver, and other exotic offerings in exchange for copper. They also brought new types of pottery and other new forms of technology. They introduced the use of large earthworks such as pyramids built of wood and earth and the burial of valued goods along with the dead at gravesites. The Hopewell evidently were not invaders who sought to conquer the Wisconsin tribes, but they had a strong influence, which ended rather mysteriously around 500 CE.

The Late Woodland period ran from 500 to about 1200 CE. As the bow and arrow came into use, hunting became more efficient. Farming villages grew, more so in the southern part of the state than in the north with its shorter growing season. People learned to store food, and for some tribes, farming became more reliable than hunting and gathering. There is some evidence of increasing conflicts among groups vying for good fishing and hunting areas, wild rice beds, and farming land.

This was the period when Wisconsin's famed effigy mounds came into use. Although some archaeologists think mound building started earlier, effigy mounds do not appear to have been associated with human burials as earlier ones might have been. Mound builders used conical and linear shapes as well as shapes of animals, including bears, birds, turtles, and panthers. While archaeologists do not know why they were built, the mounds might have served as territorial markers, clan markers, or maps.

The last period of Native American history preceding the immigration of Europeans was the Mississippian, running from 900 to 1600 CE. The people continued hunting and gathering but the farming of corn, beans, and squash became more established. Some scholars refer to the later part of this period as the Oneota tradition. It is characterized by more permanent villages located on lakeshores and in river valleys, mostly in southern Wisconsin. Gardening was a strong part of this tradition, and some villages contained cemeteries.

Many of Wisconsin's state parks also reflect the later history of the state, including the arrival of European explorers, fur traders, and missionaries, the massive immigration of European settlers, and the economic activities that followed—farming, commercial fishing, logging, and much more. Some of these stories are included in the chapters of this book, but because I have focused primarily on geology and natural history, along with some early human history, this timeline ends here.

2.1 Big Manitou Falls, Pattison State Park

2

The Rift Zone

The Northwest Corner

The northwest corner of Wisconsin was shaped by unimaginable forces of extreme heat and cold: heat so intense that it melted rock deep in the Earth, forced it to the surface, and forged massive new layers of rock thousands of feet thick over much of the area, and cold so intense that it defied the sun, captured and froze the Earth's water for centuries, and built a crushing deep field of ice that spread over most of the state. The interactions of these great masses of rock and ice were largely what created the fascinating landscape of northwestern Wisconsin (Figure 2.2).

About 1,900 million years ago, the area was near the south coast of the ancient Lake Superior continent, which was located just south of the equator. If you could go back in time and stand on that shore, you might see a row of volcanoes smoldering offshore. You might also feel the Earth rumbling because the continental plate bearing those volcanoes would be in the process of colliding with the continent under your feet. Earthquakes would be frequent.

About 85 million years later, this area was on the northern edge of a mountain range, pushed up by the collision of the Superior continent, those volcanoes, and a small continent to the south of them. The resulting Penokean Mountains stretched across an area comparable to that of today's Sierra Nevadas and might have been at least as high as the highest peaks in that range. A barren hot highland of rock and sand, the continent was devoid of plants and animals. Lake Superior was nowhere in sight, but the mountain range was flanked by a lowland to the north, the future site of the great lake.

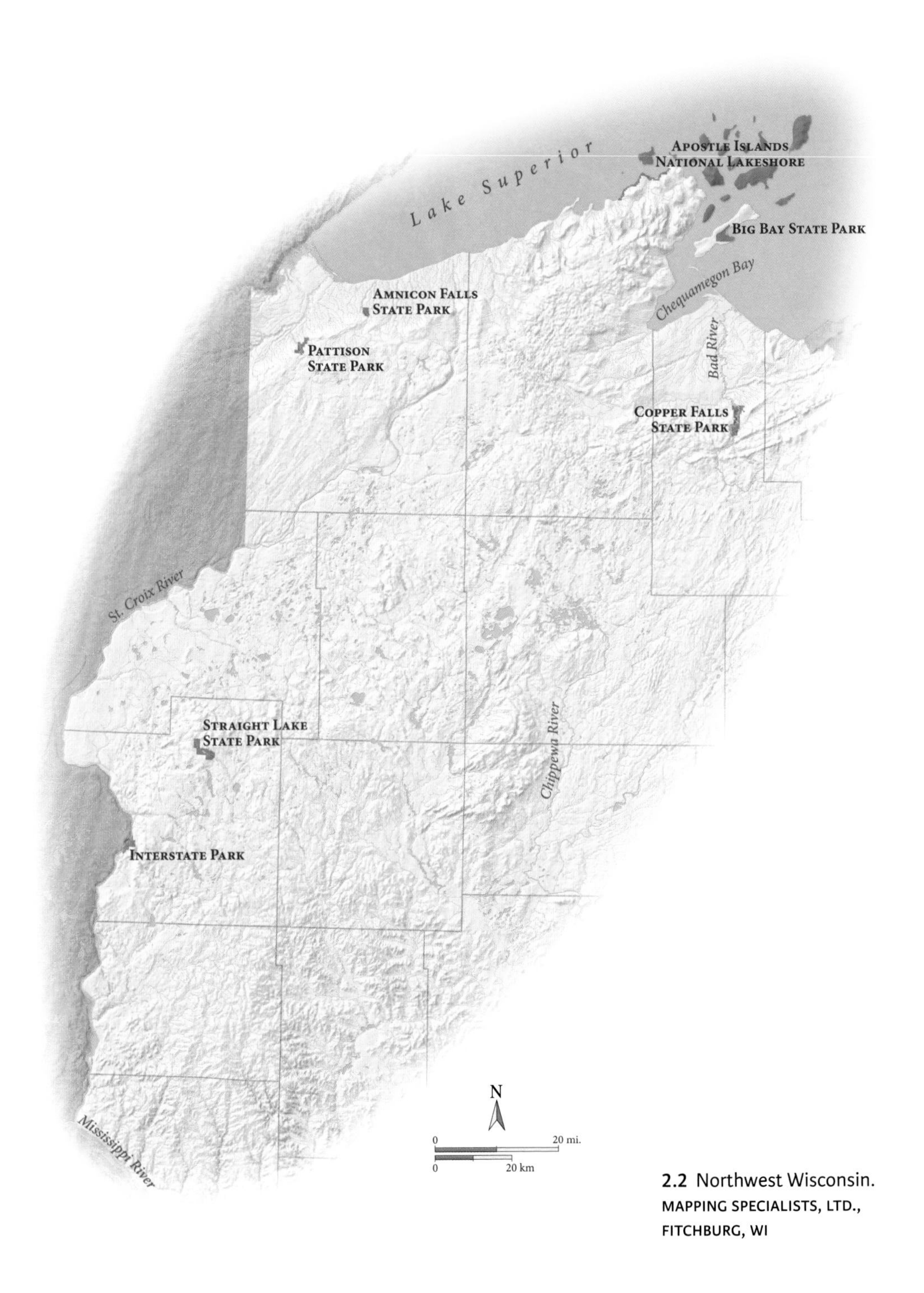

2.2 Northwest Wisconsin.
MAPPING SPECIALISTS, LTD., FITCHBURG, WI

With no plant cover, the rocky land eroded quickly and severely when whipped by warm winds and drenched by rainstorms. Within 500 million years, these erosive forces had dismantled the mountains and leveled the land to a gently rolling tropical desert area.

About 1,109 million years ago, this lifeless desert appeared quiet most of the time, but deep underground, beneath the central axis of what was to be the Lake Superior basin, a huge plume of magma was rising into the crust, fracturing it and prying it open. This slow-motion explosion signaled the beginning of Keweenawan time and eventually caused a rift that threatened to split the continent in two. As it was stretched and pulled apart, the land above the rift slowly sank, becoming a vast trough, or syncline, that contained most of today's Lake Superior area. Over time, streams washed sand and gravel into this lowland, creating shallow lakes, and subsidence, the sinking of the land, continued.

Millions of years after this process began, the rising plume reached the surface through a web of fissures in the crust and spurted lava onto the land. The lava that flowed through these fissures erupted gently, probably resembling fountains rising out of cracks in the lowland. A few volcanoes were also scattered in the rift area, and their eruptions were likely much more explosive. The cooling lava from the slower eruptions became the dark, heavy basalt that made up most of the newly formed rock. Lava from explosive eruptions formed the lighter, more granitelike rhyolite, which is less common but plentiful in the rock record of the area.

The slower eruptions continued for 25 million years, building a mass of basalt, layer by layer, over the Lake Superior region. With all of this added weight, subsidence in the lake basin continued, while sediments and lava rock filled that void as it deepened. Geologists have estimated the average depths of the various rock layers in the Wisconsin area of the Lake Superior basin. The bottom layer was 300 to 400 feet of sandstone and conglomerate, covered by about 20,000 feet of basalt. The sedimentary layers over the basalt mass eventually became as deep as 22,000 feet in some areas.[1]

At the end of Keweenawan time, once the rifting and volcanic activity stopped, the land was quiet for several million years. Then, about 900 million years ago, this period of stability came to an end as another continent collided with North America at its southeastern edge, hundreds of miles from the rift zone. The resulting compression caused blocks of land (fault blocks) to be thrust upward, creating landforms called horsts (Figure 2.3).

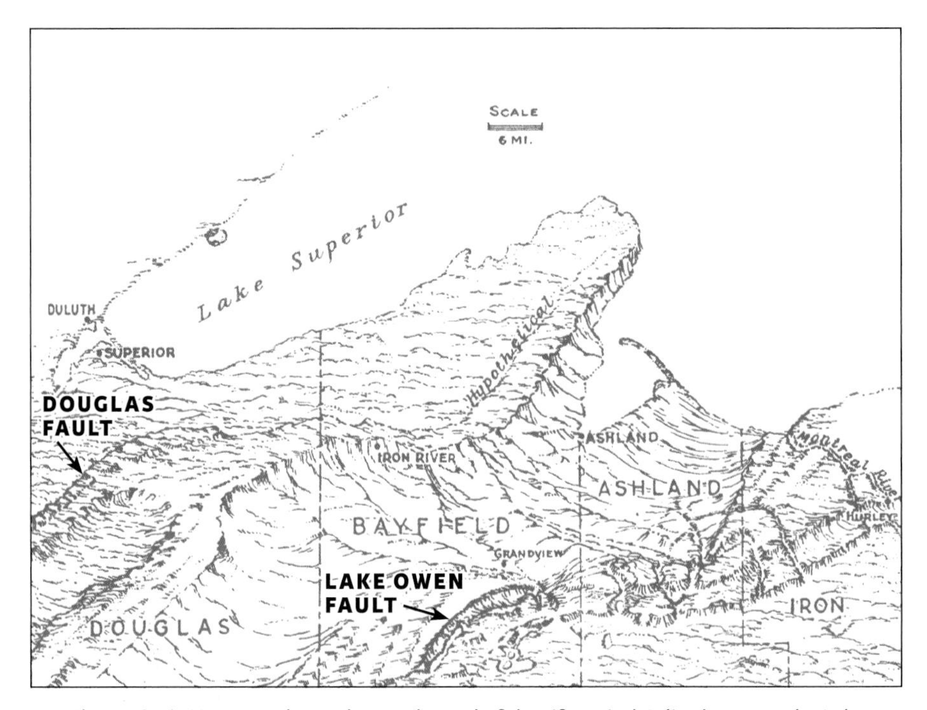

2.3 The St. Croix Horst was heaved up at the end of the rift period. It lies between the Lake Owen Fault (bottom center) and the Douglas Fault (center left). ADAPTED FROM WISCONSIN GEOLOGICAL AND NATURAL HISTORY SURVEY

One such fault block is bounded by roughly parallel lines running southwest to northeast within the rift zone. Named the St. Croix Horst, after the St. Croix River, which flows along much of its length, it is bounded on the southeast by the Lake Owen Fault and on the northwest by the Douglas Fault. Both faults played key roles in the formation of some of the parks covered in this chapter.[2]

Over about 10 million years the horst rose, squeezed between the masses of crust on either side of it. Geologists estimate that on some stretches of the Douglas Fault, it rose many thousands of feet—so high that the layers of basalt buried beneath the deep layer of sandstone were lifted above the sandstone covering the adjacent land. As it rose, the land south of the fault jutted sharply over the sandstone to the north at 55 to 60 degrees from horizontal (Figure 2.4). The fault heaved up slowly—a few inches per century—and erosion wore away the rising ridge at the same time. This kept the horst from getting very high above the surrounding plain. Today, it is level with the land around it, except in a few places where it is still prominent.

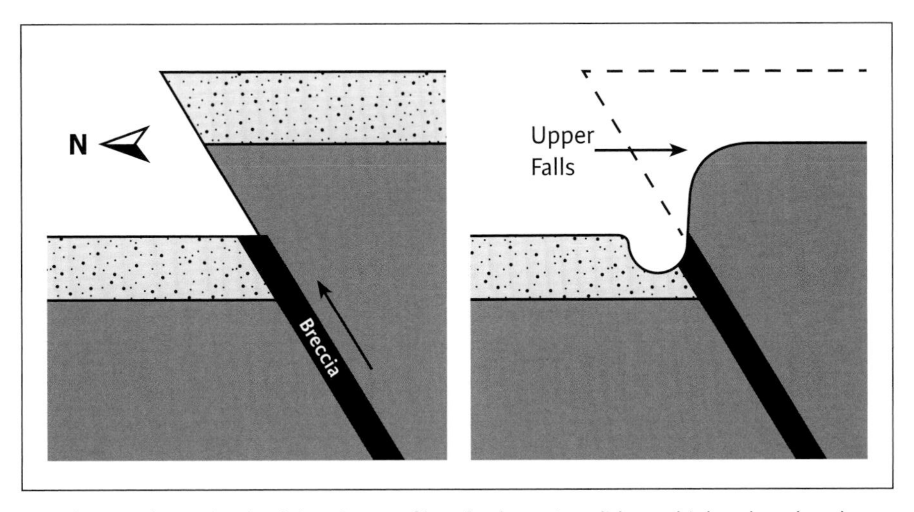

2.4 The Douglas Fault raised deep layers of basalt, shown in solid gray, higher than the adjoining level of much younger sandstone. On the right, see the long-term effects of erosion.
AMNICON FALLS STATE PARK

Faulting action fractured the surface sandstone layers at the Douglas Fault. As the south side heaved up, it lifted some of the sandstone beds on the fault's north side to a nearly vertical position. The mashing of rock against rock pulverized the sandstone in the fault zone and broke the basalt into shards. Over centuries, with the help of heat and pressure, the sharp-angled basalt fragments were cemented together within a mix of sediments. The result was a form of conglomerate rock called breccia (Figure 2.4).

The slow continental compression and upthrusting of land eventually came to an end in northwestern Wisconsin. During the last several hundred million years of Precambrian time, erosion leveled highlands and covered the area with sediments that would become new layers of sedimentary rock. Later, Cambrian and Ordovician seas probably lapped at the southern part of the state's northwest corner. The area was probably under water during the Silurian period, but possibly not during the Devonian period. Due to erosion, no rock record exists in this part of the state for the roughly 400 million years between early Devonian time and the Quaternary period.

When the Ice Age arrived, Lake Superior did not exist. Its basin was shallow at the time, mostly filled with deep layers of sedimentary rock over equally deep lava rock layers. It was probably drained by a river system flowing northeast. When the ice mass came from the north, the shallow basin channeled part of it to the southwest. Over nearly 2 million years, succeeding glaciers moved much of the basin's sedimentary rock to the southwest. When the last advance of the Wisconsin glacier approached the southwest end of the basin, it ran into the high mass of sandstone that had accumulated there as the raw material from which the Bayfield Peninsula and the Apostle Islands would be formed (see Figure 2.2). In this way, the several advances of ice deepened the lake basin and sculpted the land that would become the Lake Superior shoreline.

One of the most important effects of glaciation is the building of moraines—ridges of glacial debris. A glacier always advances to a line where it can go no farther, usually because ice at the front edge is melting as fast as the glacier can push more ice to that margin. The front wall of the glacier might then sit in that position for hundreds or thousands of years while clay, sand, gravel, and boulders are hauled by moving ice as on a giant conveyor belt to the margin. There it is dropped and accumulates as a moraine.

Even as it retreated, the Wisconsin glacier affected the land dramatically. Much of the material plowed out of the lake basin went into the moraine that meanders across northern Douglas, Bayfield, Ashland, and Iron Counties. Some of this till was also left atop the high ridge of Bayfield Peninsula. Another glacial mark on that peninsula was the enormous pile of sandy outwash, the material deposited by broad streams flowing away from the melting ice sheet.

As the ice shrank northward, its meltwater formed lakes in the lowlands, including two on either side of the Bayfield Peninsula. As they grew they merged into one large body called Glacial Lake Duluth, the ancestor of Lake Superior. Its northeastern end was blocked by the retreating wall of ice. At its southwestern end, it rose against frigid, rocky shores. Over hundreds of years, the lake level rose and fell. It got as high as 500 feet above Lake Superior's present level (which is 600 feet above sea level). At that point it lapped against the moraine in the northern tier counties. On the Lake Superior lowland, lying between this moraine and the present-day shoreline, the glacial lake deposited the heavy reddish clay, silt, and sand that are so common along today's south shore.

The glacier had left other new landforms atop the bedrock, including hummocky terrain—random arrangements of hills, ridges, ravines, hollows, lakes, and wetlands. It was formed by the uneven collapse of layers of supraglacial (top-of-the-glacier) debris as the ice beneath melted away. The hills and similar higher landforms are called hummocks. Many of the hollows, lakes, and wetlands occupy kettles, or depressions in the Earth formed from remnant blocks of ice that were surrounded or buried by the debris washing away from the glacier as it retreated. When these remnant ice blocks finally melted, in some cases after hundreds of years, they left depressions.

Another prominent postglacial landform was the tunnel channel—a long, linear valley carved by a fast-flowing stream that flowed under and out from the base of an ice mass as it melted back. These subglacial streams often carried debris and dropped it along their beds, creating curvilinear ridges called eskers. These glacial landscape features appear all over northern Wisconsin and are showcased in some of this chapter's parks. Chapter 5 covers glacial dynamics and features in more detail.

As the glacier shrank away, ice-free areas gradually warmed, becoming more hospitable for forests, plant life, and eventually humans. Archaeologists think the first humans to arrive in northern Wisconsin were Early Archaic Indians carrying on the Paleo-Indian hunting tradition in pursuit of mastodons and other big game about 9,000 years ago. Over the succeeding centuries, Late Archaic and Woodland peoples hunted and fished along the riverbanks and lakeshores and gathered plants from the forests between 5,000 and 1,500 years ago. The tribe most prominent in the area when Europeans began migrating were the Ojibwe (also called the Chippewa). While natives in southern Wisconsin were developing agriculture on land leveled by the glacier, people of the north were faced with a shorter growing season. They did some gardening but tended to rely more on fishing during summer and hunting year-round for their food sources.

Natives also mined copper, mostly in Michigan's Upper Peninsula, but some in northern Wisconsin as well. After the glacier unearthed deposits of copper, it could be picked up from the ground and dug from shallow deposits. The Indians never smelted copper, as Europeans did later, but instead hammered the copper into various shapes to make spear points, knives, awls and other tools, ornaments, and other goods.

Northern Wisconsin provided fertile hunting grounds for the native peoples who lived here sustainably for thousands of years. Not so sustainable was the later use of the great pine forest that had matured in northern Wisconsin long after the glacier was gone. Europeans migrating to North America viewed the forest—often referred to as "the pinery"—as a vast resource to be harvested. Harvest they did, clearing most of the state of its old-growth pines between the 1830s and the early twentieth century.

The glaciers pulverized much of the evidence of bygone millennia and left a layer of glacial till over most of northern Wisconsin. In areas of the parks we will visit, this covering was worn away by thousands of years of erosion. This and other postglacial effects resulted in the features that make many of the state parks unique and fascinating.

Pattison State Park

A few miles south of the Lake Superior shoreline, an east-west-trending range of high hills rises from the lowland that borders Lake Superior (Figure 2.2). They were built on a foundation formed by the Douglas Fault. For centuries, the glaciers transported sand, gravel, and rock from the north to build the broad highland.

Visitors approach these hills while traveling south on State Highway 35 from the city of Superior across the lakeshore lowlands. Near the top of the first rise is Pattison State Park, which straddles the ascent from the lowlands to the highlands. The most popular parts of the park are in the highlands, and the most visited site of all is Wisconsin's highest waterfall—and the fourth highest east of the Rocky Mountains—Big Manitou Falls (Figure 2.1).

The 165-foot waterfall is part of a spectacular gorge created by the Black River, which originates southwest of the park. In the park area, the river encounters the basalt mass exposed by the Douglas Fault. It plunges over one basalt ledge at Little Manitou Falls, flows gently through Interfalls Lake, and then enters a cascade that leads to Big Manitou Falls.

Between the time of the Douglas Fault and glacial times, the fault area was more rugged than it is now, although wind, rain, and ancient seas had eroded the land to a gently rolling plain. The glaciers that covered the area several times did more scouring and leveling of the land. As the last glacier retreated, Glacial Lake Duluth covered most of the park area, and remnants of its beaches can still be found.

As the glacial lake drained, it left a clay-covered slope on its south shore. The area remained frigid and great blocks of remnant ice and bodies of dammed ice water lay in the highlands to the south. As temperatures rose, this stored water was released through streams flowing north off the highlands. Geologists think some of these flows, including that of the Black River, must have been rapid. The river quickly cut through the clayey cover to the underlying basalt south of the fault. North of the fault, it wore away at the sandstone, carving a channel 100 to 170 feet deep downstream from the falls.

In the fault area, the basalt was much harder for the water to carve, but the river is thought to have coursed through a cross fault (a smaller fault lying at a sharp angle across the main fault, caused by the violent fracturing of the crust). In such a fault would be a breccia zone, which was much easier for the river to cut through. Over centuries, the river worked at the basalt and breccia, slowly carving the gorge now containing the falls. Visitors can see that much of the gorge lies within the basalt mass—that the river plunges down several hundred feet upstream of the fault line.

The Ojibwe, who had arrived in the park area long before Europeans came, were at least as enthralled with the waterfalls as today's tourists are. In the roar of the big falls, they heard the voice of Gitchi Monido, Ojibwe for "Great Spirit." They revered the falls, making it a central gathering place for meetings and ceremonies. The name Manitou is perhaps a French fur trader's version of Monido.

Some Native Americans picked up copper in the park area and dug shallow mines to find it, especially near Little Manitou Falls and in the far northeastern part of the park near Copper Creek. They hammered copper into tools and ornaments and perhaps traded it for goods from other tribes. They also valued quartz, which they dug from the conglomerate rock downstream from the big falls. They chipped and carved this very hard rock to make spear points, arrowheads, knives, and tools.

European Americans also came to the park area in search of copper starting in about 1845. Miners dug shallow pits at the bases of both waterfalls and in the Copper Creek area of the park. Because of rough terrain and other challenges, they had little success. Miners returned several times over the next 50 years to try to make a go of it in the park area. The deposits found were seldom rich enough to make it worthwhile.

The park area was exploited by the lumbering industry of the late nineteenth and early twentieth centuries. One young lumberjack was Martin Pattison, who worked first in Michigan and soon owned a lumber company. In the early 1880s, his company logged the slopes along the Black River. Pattison grew wealthy due to his success in lumbering and later in iron mining in Minnesota. In 1917, he got wind of plans to build a hydroelectric dam on the Black River—a structure that would have drowned Big Manitou Falls. While working in the area years earlier, he had fallen in love with the gorge, so he used his considerable wealth and power to purchase the land to preserve it. Within a year, he had donated his purchase to the state, and in 1920, it was established as Wisconsin's sixth state park, named in his honor.

Pattison State Park became a popular destination and, closer to its natural state, was more primitive than it is today. In the 1930s, the Civilian Conservation Corps (CCC) did a tremendous amount of work to preserve the park's natural features in the face of growing numbers of visitors while also making the park more visitor friendly.

The CCC workers established Camp Pattison off Highway 35 near Little Manitou Falls, in what is now the picnic area. Between 1934 and 1942, they built the shelter building and bathhouse from sandstone blocks. Both buildings still stand, with the shelter now used as a nature center. The CCC workers also put in sewer and water systems, landscaped the lakeside recreational area, planted trees, and hauled many tons of sand to create the park's beach. They helped to build the campground and pedestrian underpass to Big Manitou Falls, and they rerouted County Highway B around the park. Finally, they built the three miles of hiking trails that now give us access to the park's most popular attractions.[3]

TRAIL GUIDE
Big Manitou and River Trails

These are easy-to-moderate trails, the hardest part being a steep descent on the River Trail. They total less than a mile, and the total change in elevation is about 170 feet. From the park headquarters, you can get a pamphlet called *Pattison State Park Big Manitou Geology Walk*, which takes you to several vantage points overlooking the falls from both sides of the gorge.

From the parking lot near the park entrance, take the short walk to the underpass under State Highway 35. Leaving the underpass and walking west (to your right), you view the cascade of the Black River before it plunges over the falls. There the river turns north, as does the trail to three interesting overlooks.

The first overlook is directly above the falls. Here you are standing in the part of the gorge carved within a cross fault. The wall below you and the wall on the opposite side are basalt. If you look closely at the walls, you might discern layers, although they are nowhere near as obvious as those in the sandstone farther down in the gorge. A basalt layer changes color, usually from blackish to reddish or from darker to lighter brown, from the bottom up. This is because the top of each flow was weathered and oxidized as it cooled, sometimes for hundreds to thousands of years, before the next flow covered it.

The next stop is built out away from the face of the gorge. Looking downstream, notice that the wall on your side of the gorge is a lighter-colored sandstone. On the opposite side (the west side), the dark-colored basalt wall still blocks the view of where it changes to sandstone. What you are seeing is the Douglas Fault—the point where the basalt mass rose against the Lake Superior sandstone overlying it. If you could turn back the clock to postglacial times, you would be standing near a cascading river flowing at about your level and you might be able to hear where the cascade became a much steeper set of falls downstream at the fault line. Since then, the river has carved the upper gorge and the waterfall has worked its way back to its present location.

At the next outlook, farther north on the trail, is the approximate location of the Douglas Fault. Standing here, you can imagine riding the land up, as if on a glacially slow elevator, as it rose above the land to the south. A wooden information sign near this point informs you that "a mighty lake"—Glacial Lake Duluth—once covered most of the park. When that lake was at its highest level, the place you are standing would have been offshore under about 100 feet of water.

From here, walk back on the trail, past the top of the falls and upstream along the river to the footbridge near the highway underpass. On the other side of the river is a boardwalk along the west rim of the upper gorge. The first overlook on this boardwalk offers a good view of the west side of the falls, although the best view is still to come. Here you can see a pit at the base of the falls made by copper miners near the end of the nineteenth century.

The boardwalk ends and the trail continues to the next overlook—the favorite viewpoint for photographers (Figure 2.1). This is also the approximate point where this trail crosses the Douglas Fault, which runs from here across the gorge to where you crossed it earlier. Under your feet is the zone where the basalt ends and the layered sandstone begins in the wall below. At this viewpoint, you can get a feel for the massive nature of the basalt from which the upper gorge was carved. The brown and black rock wall represents just a fraction of the total depth of this lava rock. Looking back in time 1 billion years, you can get a sense of how long it must have taken for these walls to be built, layer upon layer, by the lava spreading from fissures to the north—an estimated 25 million years.

2.5 Layers of sandstone on the bank of the Black River below Big Manitou Falls.

You can also begin to appreciate the persistence of water working on rock. The combination of fast-flowing water in summer and frost in winter has gradually moved this waterfall back from the fault line under you to where it is now. Try to imagine how long it takes the rushing water to scrub away just a fraction of an inch of this rock and you might get a sense of how long that recession of the falls took. High up in the gorge near the falls, some sections of the wall have a rust-colored hue. These are patches of lichens, a hardy combination of algae and fungi that are helping to weaken and dismantle the basalt walls.

Down the trail at the next overlook, you can view the rapids below the falls, spectacular in their own right. The Ojibwe called these cascades Bohiwin-Sasiqewon, which means "laughing rapids." Here as in the falls, notice the root beer color of the river water. It is caused by tannins released into the water by decaying vegetation in the vast bogs and swamps of the upper stretches of the Black River.

From here, the River Trail descends into the lower gorge, carved from sandstone north of the fault. It first winds gently down through piney woods for about a third of a mile, after which it becomes moderately steep, dropping down to the river. About a half mile past where you viewed the rapids, you descend a set of wooden stairs down the face of what seems to be a small escarpment or ridge, about 15 feet high, running through the woods as far as you can see in either direction. This is an ancient riverbank built of sand and gravel moved by the water in a time when the glacial lake was much higher than present-day Lake Superior. The higher lake level would have made the river much wider and deeper than it is now in this lowland below the falls. Higher up on the big escarpment you just descended, geologists have found evidence of several ancient beach areas left by the glacial lake as it slowly dropped to lower levels.

About 100 yards farther down, the trail ends at the riverside, and you get a good look at the sandstone beds that make up the lower gorge wall (Figure 2.5). The river is quiet here, flowing lazily now as it begins its meander toward the Nemadji River and Lake Superior. Here again, take some time to imagine how long it took for the sandstone wall across the river to be built.

TRAIL GUIDE

Beaver Slide and Little Manitou Falls Trails

Beaver Slide is a two-mile loop trail, and the trail to Little Manitou Falls from Beaver Slide Trail is another half mile. The hike from the parking lot to Little Manitou Falls and back is three miles. The full loop is an easy-to-moderate hike with a few short ups and downs. Although Little Manitou Falls can also be reached by driving south from the park entrance on Highway 35, the hike is well worth the effort.

From its trailhead on the east side of Interfalls Lake near the bathhouse and beach, the Beaver Slide is a flat trail through a lowland next to the lake. On some stretches, the woods are dominated by spruce and cedar, much as they have been for thousands of years, since the last glacier retreated.

A little over a half mile from the trailhead, you have passed the lake and are walking upstream on the Black River. Near the one-mile mark, you cross the river on a bridge where the Little Falls Trail continues upstream. On the hike to the falls, you will see sandy areas on the trail and in the woods—remnants of an ancient beach. Somewhere downstream from Little Manitou Falls was the south shore of Glacial Lake Duluth at its high point. Turn around and imagine yourself on that shore, looking across a vast expanse of frigid water. You might see icebergs floating far out, having calved from the glacier's retreating wall of ice to the northeast. Behind you would be a lifeless sloping expanse of wet, rocky land, speckled with patches of ancient ice—a newly built moraine of the departed glacier.

About a half mile from the trailhead, you take a short, moderately steep climb on blacktop to good views of the falls. Here the river drops 31 feet over a basalt ledge into a breccia zone. You are more than a mile south of the Douglas Fault, but this reminds us that a major fault zone comprises more than just one big crack in the Earth. It includes a complex pattern of cracks on either side of the main fault, some of them running not with the fault line but at angles to it. Little Manitou Falls drops into one of these minor faults.

Returning on the Little Manitou Falls spur trail, you come to the junction with Beaver Slide. The section of the trail around the west side of the lake is quite different from the level east-side section. It crosses hilly, wooded terrain where diverse native vegetation is holding its own. The flora here is much the same as what the Ojibwe would have seen in the days before this area was taken over and developed by European immigrants.

About three miles from the Beaver Slide trailhead, the trail leaves the woods and arrives at the north shore of Interfalls Lake. To get back to the beach and parking area from here, you must cross the grassy area between lake and road and then cross the metal gridwork walkway over the dam and across the river.

Amnicon Falls State Park

From the Amnicon Falls State Park entrance, the main road goes north, or downstream, along the river for a quarter mile to a parking area. There visitors get their first glimpse of the Upper Falls, where the river tumbles over a 30-foot rock wall (Figure 2.6). It is one of those rare places where the Earth has opened up to give us a view to the distant past.

The Amnicon River originates in northern Douglas County and flows just 30 miles, descending 640 feet to the south shore of Lake Superior. About midway on its course, it flows for two miles through Amnicon Falls State Park. In that stretch, the river drops 180 feet over waterfalls that are spectacular when flows are high. Three of the park's four main waterfalls flow over the Douglas Fault, which lies at the center of the park's geologic story.

2.6 Upper Falls, Amnicon Falls State Park.

Some geologists estimate that after the Douglas Fault formed, had it not been steadily eroded, the escarpment on the south side of the fault in the park would stand 15,000 feet or higher above the land to the north.[4] Today it is just about 30 feet higher at Amnicon Falls.

After the formation of the fault, the next known major event to leave its mark in the area was the advance of the ice sheets during the Quaternary period. In and around the Amnicon River bed, glacial till has been removed by flowing water and ice, but glacial erratics, or boulders carried from the north and dropped by the glacier, are present. When the ice melted, the river carried huge volumes of water and ice chunks. Where the flowing water formed shallow eddies, it captured small pieces of basalt or harder rocks and spun them ferociously against the basalt riverbed. Over many years, this spinning action wore kettle-shaped holes called potholes in the bedrock. Such torrents of water also washed away large volumes of sandstone in the park area, taking it downstream toward Lake Superior.

As the glacier finally retreated and forests grew on the upland stretches of the Amnicon, Archaic and later Woodland and Ojibwe Indians used the resources provided by the forest, wetlands, river, and lake. The name Amnicon is derived from an Ojibwe word meaning "where fish spawn." The mouth of the river has always been and remains an important spawning area for Lake Superior fish. The park's Thimbleberry Nature Trail, located on the west side of the river downstream from the falls, is named for a plant that produces tart, dry berries related to raspberries. Natives used them for flavoring and also cooked and mashed them to make small cakes that could be stored and eaten during winter.

The first Europeans to frequent the banks of the Amnicon River were trappers after beaver, river otter, and mink. The Ojibwe also trapped these animals, but only for sustenance. European trappers sold the furs for profit. While in competition, these two groups cooperated, bartering for each other's favors and goods. Later came miners looking for copper, having heard the legends of early copper mining. Some were successful in this area, but most were not.

As part of the Treaty of 1842, which resolved disputes over lands claimed by both the United States and Britain, the land surrounding today's park became US federal land. When Wisconsin became a state in 1848, the US government gave the state much of the land in northern Wisconsin. The state in turn gave much of the land to a railroad company to spur the development of rail service in the area. The company retained its right-of-way but sold most of the land to raise money for building the railroad. One of the buyers was James Bardon, resident of the town of Superior, who in 1886 bought the land that includes Amnicon Falls State Park.

In the late 1800s, loggers came to the area to cut the pine. The Amnicon served as a conduit for the logs, which were floated down to sawmills in Superior and Duluth. The thick layers of a harder variety of sandstone in the region were quarried in and around the park area. The reddish-brown stone was a popular building material called brownstone that makes up the walls of many stout buildings in northern Wisconsin towns and in several cities, including Chicago, St. Paul, Minneapolis, and New York City. You can view a long-abandoned sandstone quarry from a spur trail off the Thimbleberry Nature Trail.

TRAIL GUIDE
Waterfalls Trail

Of the park's three trails, Waterfalls Trail makes the best introduction to the geologic and natural history of the park area. I walked this trail in early June when the water level was high and the falls were roaring. For much of the year, they are not as spectacular, but when the water is lower, you can more easily see some of the features that tell the geologic story of the park. This is an easy, almost level trail. Be aware that it lies very near the edges of some steep drops into the river, especially near Snake Pit Falls.[5]

From the parking lot next to the falls, walk to the vantage point just to the left of the covered bridge and view the Upper Falls upstream to your left. The river flows over dark reddish-brown rock—basalt formed from the lava that erupted in the Lake Superior basin starting about 1.1 billion years ago.

Continue upstream and down the steps on the riverbank to the viewing platform nearer Upper Falls. From here you can look straight at the falls and study the feature that makes it possible—the Douglas Fault. Compare the solid-looking mass of basalt over which the water flows with the lighter-colored and layered sandstone downstream. Note that, over millennia, the river carved a notch in the basalt wall and the falls has migrated a short distance upstream.

Across the river and just downstream from the falls is the breccia zone, the jumbled fragments of basalt cemented together to form a layer about 12 feet thick. Further downstream are horizontal sandstone layers, but notice that at the fault, those layers were heaved up almost to a vertical position by the upward thrust of the adjoining basalt. The water here is reddish-brown, often described as root beer colored. This is because the river originates in bogs and conifer swamps to the south. As vegetation in these wetlands decays, it produces tannins, which stain the river's water.

As you climb the steps and turn downstream, it is a short walk past the covered bridge to the flat overlook of the river downstream from Lower Falls. These falls, also about 30 feet high, drop over a precipice made of Precambrian sandstone. Scan the riverbanks downstream from these falls and observe that they consist of several layers of sandstone, also called bedding planes, that took millions of years to form. Collectively, they are called Lake Superior sandstone and are thought to be about 3,000 feet thick in the park area.

Underfoot, notice the cracks that run across the platform of rock. They provide a cross-section view of vertical planes called joints—cracks formed over the centuries by seeping water that freezes and expands during winter. The river continually wears away at the wall on which you are standing. Eventually it erodes enough stone at the river level, low on the wall, and the weight of the overlying stone pulls a mass of stone away from the larger body of sandstone at one of the joints. That mass of stone plunges into the river and leaves a sheer rock face on the wall. In this way, the river has widened its bed here below the falls.

Now, walk back to the covered bridge and cross the river to the island lying between two branches of the Amnicon River. If you turn left and walk upstream, you can get another view of Upper Falls and the Douglas Fault. Continuing upstream toward the falls, you pass a low berm of conglomerate near the riverbank. The rounded stones and pebbles within this formation are evidence that this was once the shoreline of the lake, where those stones and pebbles were tumbled by waves for thousands of years. They were later covered by sandstone and cemented into conglomerate long before erosive forces exposed them again.

Passing the falls and continuing upstream, you can look across the river to see excellent examples of glacial erratics—boulders transported from as far away as Canada. They are made of granite and other types of rock not common to the area, which is why we know they were moved here from the north.[6]

At the point where the river branches, crossing the footbridge that spans the west branch of the river takes you to the campground, from which the Thimbleberry Nature Trail runs to the right of the bridge. I turned left off the footbridge and hiked upstream on the west bank of the river, above where it branches, and I recommend this side trip. You can get a view of the rapids above the falls and stop at one of the several picnic tables for a quiet rest or a lunch. Pines, cedars, and spruce dominate the woods beyond the trail. The smell of pitch is strong, and the drooping trees, moist green hollows, and scattering of moss-covered boulders make the woods seem ancient—as if you have strayed into a forest somewhere in Tolkien's Middle-earth.

Back across the footbridge on the island, it is a short walk downstream along the west branch to where you can view Snake Pit Falls. Like Upper Falls, this one also flows over the fault. When the river is high, this is a roaring cataract that plunges through a series of deep cauldrons. At the points closest to the falls, it is a sheer drop into the rushing river, so hang onto children and pets in this area.

2.7 This wall of bedded sandstone just downstream from Lower Falls is being eroded back by the Amnicon River.

Just beyond Snake Pit Falls near the northwest corner of the island, you walk along a high rock ridge overlooking the river. The steep ravine is carved from the breccia zone within the fault. On the island, the glacier and other forces have leveled the land so that you walk across the fault without knowing it. However, as erosion continues in centuries to come, this ravine will grow, crossing the trail and splitting the island along the fault zone.

Keep following the trail around the perimeter of the island and you will pass the place where the branches of the river rejoin at a junction that is out of sight from the trail. You will now see the north side of the island and, across the river, get a spectacular view of the sandstone layers that make up the wall on that side of the river. Notice how the river is eroding the wall from the water level up (Figure 2.7). Someday another chunk of that wall will drop into the river, leaving a new sheer wall of sandstone.

Crossing back over the covered bridge, walk east across the parking lot for a view of Now and Then Falls. The stream is yet another small branch of the Amnicon that flows over the Douglas Fault. Even during high water, this is a much smaller falls with a more delicate sort of beauty. In drier times, it does not run at all, hence its name. Your vantage point is located on a mass of conglomerate and after the stream drops over the fault, it veers east along this wall of conglomerate and erodes the breccia within the fault zone. This ravine will continue eroding westward and some day could join with the ravine near Snake Pit Falls that is eroding eastward along the fault.

Big Bay State Park

From the air, on a summer day, the Apostle Islands look like jewels scattered in the waters around the Bayfield Peninsula north of Chequamegon Bay (Figure 2.2). Madeline Island is the largest and most developed of the 22 islands. Some have likened the shape of the island to a side view of a crocodile with its mouth open. The oversized snout of the crocodile is pointed northeast, and its short lower jaw and mouth are aimed straight east, away from the peninsula. The croc's mouth is called Big Bay, the site of Big Bay State Park.

The story of Big Bay begins with the story of how the Apostle Islands were formed. The Bayfield Peninsula and the islands lie on or near the axis of the Lake Superior basin, the line running southwest to northeast along its lowest level. This is where the land that was to become the lake bed subsided to its lowest point during and after the Midcontinent Rift event around a billion years ago. Between then and 500 million years ago, this deepest part of the basin was filled by more than 20,000 feet of Precambrian sediments brought by streams flowing into the lowland. The resulting deep beds of sandstone were the rock from which the peninsula and islands would be formed.

The sandstone beds lay relatively undisturbed for hundreds of millions of years as lakes formed in the basin and plant and animal life took hold. When glaciers started advancing from the northeast, the area was probably a gently sloping plain covered by a boreal forest and drained by a river running northeast roughly along the axis of the basin.

The glaciers that advanced multiple times over nearly 2 million years gradually carved the lake basin and sculpted the great mass of Precambrian sandstone at its southwest end. This sandstone mountain divided the last advance of the most recent glacier into the Superior and Chippewa lobes. The two lobes moved along its northwestern and southeastern flanks, respectively, while chewing away at its northeast end, as previous glaciers had done.

The Apostle Islands were probably part of the peninsula at one time. The action of the ice masses carved out some of the low-lying areas that separate the islands from the peninsula and from each other. The peninsula has high hills on its eastern side, separated by stream valleys. Historical geologist Robert H. Dott Jr. has suggested that these valleys might have been channels eroded beneath the ice mass by meltwater that washed away more of the sandstone over thousands of years and left the hills much as they are today. Dott sees the Apostle Islands as "the partly submerged continuation of these sandstone hills."[7]

As the glacier retreated, Glacial Lake Duluth covered the islands for many centuries, leaving a layer of red clay on the islands and much of the peninsula. Only the highest, southern portion of the peninsula was above water when the glacial lake was at its highest—a large sandstone island near the south shore of the lake.

After the glacier was gone, Lake Superior continued to shape the islands. They are famous for their spectacular red-and-brown sandstone cliffs and caves carved by wave action over the centuries. At Big Bay State Park, visitors can study such cliffs on the Bay View and Point Trails (Figure 2.8). The sandstone on this and other island shorelines lies in distinct layers, or beds, some of which lie at angles to others—a phenomenon called crossbedding. Such angular bedding often indicates a place where the sand was laid down on a slope, such as a steeper shore or a sand dune. Geologists estimate that these sandstone beds have a total thickness of as much as 1,000 feet.

Another example of the lake's shaping influence is the barrier beach at Big Bay State Park, one of the park's most popular attractions (Figure 2.9). When the glacial lake drained to Lake Superior's present level, the bay was much bigger, occupying what is now the park's lagoon and beyond. Without the rounded shape it has now, the bay would have resembled a crocodile's mouth even more, jutting westward into the island's interior.

On certain types of sandy, gently sloped shorelines, wind, water currents, waves, and the shape of the land work together to build sandbars parallel to the shore. Over many years, they grow to the point where they become barrier beaches, or long, narrow islands along the shore. This has happened at least

2.8 Bedded sandstone underlying Big Bay State Park is exposed all along the Bay View and Point Trails.

twice at Big Bay. Earlier in postglacial times, a barrier island was built farther west in what was then the bay. The water between that island and the shore gradually filled with sediments and plants and is now part of an extensive wetland that overtook the older barrier beach. Later, the present barrier beach was built farther out in the bay. The lagoon that now lies between the wetland and the younger barrier beach (Figure 2.10) is being filled in by natural processes, just as the first lagoon was.

The lagoon and the barrier beach are both fragile ecosystems that took centuries to become established. The park is a showcase for plants that grow in areas such as sand spits where conditions change dramatically and frequently. For example, 11 or more species of ferns are common to the park. Ferns are an ancient type of plant that existed hundreds of millions of years ago before flowering plants evolved. Some of the earliest forms of plant life in Wisconsin were tree ferns that grew to over 100 feet tall. Of course, no such ferns, or any plants at all, grew in the park area until well after the glaciers created it.

2.9 The barrier beach at Big Bay State Park is the site of rare and fragile plant communities.

The plant communities of Big Bay State Park were vital to the first people to live in the area on any sort of permanent basis, starting about 1,200 years ago. They followed the Late Woodland tradition, living in temporary camps for fishing, hunting, and gathering wild plants. By the mid-1600s, when French explorers first arrived, the dominant tribe in the area were the Ojibwe, who did some trading with the French, exchanging furs and other native items for goods made of glass, metal, and cloth. From the late 1600s through much of the next century, the Ojibwe had a large village on Madeline Island, which had become a spiritual center for them.

By 1693, the French had established a trading post on the island. From that point forward, European economic activity ebbed and flowed on the islands, depending on factors such as the European economy and wars between the French and the English, and Madeline Island would see rising and falling populations of both Europeans and Ojibwe. In treaties signed with the US government in 1842 and 1854, the Ojibwe ceded their lands on the southwestern shore of Lake Superior, including the islands, and most of them moved onto the Bad River reservation on the south shore of Chequamegon Bay and the Red Cliff reservation on Bayfield Peninsula.

2.10 The lagoon that lies behind the barrier beach at Big Bay State Park.

Economic activities among European immigrants and the Ojibwe from the late 1600s on included fur trading, commercial fishing, logging, quarrying of sandstone, farming, and finally tourism, which dominates the area's economy today. The sandstone quarrying was a relatively short-lived but robust business that, during the late 1880s, furnished builders in several American cities and towns with the attractive reddish-brown sandstone of the area, often called brownstone. Those quarrying operations were mining the legacy of ancient rivers and seas that moved sand to this area over hundreds of millions of years in Precambrian time. The stone from the Apostle Islands, making up the walls of many buildings standing today in Washburn, Ashland, Milwaukee, and Chicago, is more than 600 million years old.

By the mid-twentieth century, especially as tourism became a major industry in northern Wisconsin, the area's natural beauty was increasingly valued more highly. The Apostle Islands had been considered and rejected as a site for a national park in the 1930s and later for a state park. In 1963, the state established Big Bay State Park as an area to be limited to low-impact activities such as hiking, camping, and sport fishing. In the later 1960s, Wisconsin's governor and later senator Gaylord Nelson spearheaded a push to make the Apostle Islands a national lakeshore (on par with a national park). All of the islands except for Madeline and Long Islands are part of the national lakeshore.

The combined bog, boreal forest, and dune ecosystem of Big Bay State Park contains one of the richest collections of native vegetation in the country. Such plant communities are fragile and shrinking in number and area, so they are becoming more ecologically important every year. In addition to being preserved within the state park, much of the Big Bay beach and lagoon area is designated as a state natural area, receiving additional protection as a place to be left essentially undisturbed.

Madeline Island is now a popular tourist destination with regular ferry service between Bayfield and La Pointe, the town located on the west end of the island. In winter, an ice road connects the two towns, and as the ice is forming in the fall or breaking up in the spring, you can get a ride on an ice boat. The Madeline Island Museum, a Wisconsin Historical Society site, houses an excellent collection of artifacts and information on all aspects of the island's rich history.

TRAIL GUIDE

Barrier Beach Trail

The Barrier Beach Trail is a boardwalk over sensitive terrain, just over a mile long. It includes several signs that give information on the ecosystem and its plant and animal life. The trail parallels the beach, with numerous spur boardwalk and sand trails that take hikers out to the beach for views of the lake.

A major highlight of this hike, not to be missed, is the observation platform built out a few yards into the lagoon and bog area. Compared to the lakeshore, this is a serene setting in a rare floating bog environment—one of the most diverse of such systems in the Lake Superior region. From the platform, you can spend some quiet time contemplating the long, slow process of bog formation. This was once part of the open bay and is now becoming a wetland with its own kind of beauty in contrast to that of the lakeshore.

The official park trail ends a little beyond this platform viewing area, a quarter mile from the border of Big Bay Town Park, which adjoins the state park. But the well-worn trail continues along the shore into the town park where visitors find more gorgeous views of the lake and lagoon (Figure 2.10).

Copper Falls State Park

In a 1938 report on Copper Falls State Park for the Wisconsin Conservation Department (now the Department of Natural Resources, or DNR), researcher G. T. Owen wrote, "The windborne music of rapids and waterfalls, now an interrupted murmur, now a deep, constant roar according to the vagaries of the breeze, is heard along practically every stream course."[8] It is easy to see why Owen was spellbound by what he saw and heard around Copper Falls State Park. Two rivers—the Bad and Tyler Forks—come together in the park to create a rich collection of cascades, waterfalls, and whirlpools. This wild mix of surface features reflects a complex geologic history and structure underlying the ancient, beautiful setting for the park.

The park area was once near the south coast of the ancient Superior continent. Later its bedrock was heaved up and churned during the Penokean mountain-building episode. Much later, after those mountains had been leveled, the area was covered by a Precambrian sea. The sea's advancing and retreating shoreline and the streams flowing into it laid down layers of sand, gravel, and mud that later became sedimentary rock layers.

Beginning 1,100 million years ago, the Midcontinent Rift sent multiple layers of lava to cover the area over a period of 25 million years. Geologists believe that sometime during the Keweenawan episode, a volcano in the park area erupted even more explosively, adding rhyolite and red lava rock to the local rock record.[9]

The next major event was the steep tilting of the entire area as the rift process was halted due to a continental collision to the southeast. When this compression of the land began, about 1,060 million years ago, the area was a rolling, sandy, and rocky plain overlying layers of sandstone, shale, and conglomerate, overlying the mass of basalt. The lateral compression of the land in the Copper Falls area heaved it up, tipping it steeply to the north and northwest.

With this upheaval, some of the layers of rock under the land slid over other layers. The grinding of layer upon layer created fault zones and masses of shredded rock that eventually became cemented together to make breccia zones. Coarsely defined ridges and valleys formed across the slope of the tilted land. Ancient waters kept flowing, forming rivers in the valleys and wearing away at the reconfigured rock structures. If you could fly over this area after the tilting and sweep your gaze from southeast to northwest, you would see a newly formed land surface made up of a cross-section of rock layers whose edges had been turned up.

The fracturing and jumbling of rock layers in and around the park could be related to the Lake Owen Fault. However, in the end, the compression, fracturing, and tilting of the land created a complex topography governed not by one major fault line, but by a number of short faults, some running parallel to the main fault and others crossing it at angles.

Hundreds of millions of years of erosion erased the rock and fossil records in the park area. We know that seas invaded the central part of the continent on several occasions, at times lapping the shores south of the park area, sometimes covering it. Dinosaurs may have roamed the valleys and ridges, but we will probably never know. Geologists estimate that streams running in present-day riverbeds in the park were carving its gorges as early as 200 million years ago.

The glaciers that began advancing 2 million years ago ground down the ridges and helped to wear away much of the rock in the area, sending it south. With each successive glaciation, meltwaters carved the Bad River gorge deeper. The Chippewa lobe of the Wisconsin glacier covered the park area until about 11,500 years ago. It left a deep layer of sandy till and as it retreated, meltwater flowing south from the glacier dropped sandy outwash. Glacial Lake Duluth generally covered the lowland to the north, and remnants of its beaches can be found in the northern end of the park. Between 11,500 and 9,500 years ago, the glacial lake dropped its signature red clay deposits across much of the park area. Professor Tom Fitz of Northland College in Ashland describes a distinctive strip of sand within the park separating the clay deposits to the north from the sandy glacial till to the south.[10]

The first humans to enter this area were those of the Paleo-Indian tradition, following the mastodons and other large mammals east and north as the glacier retreated. People of the Late Archaic traditions, the so-called Old Copper culture, used copper they found in the park to make tools, weapons, and ornaments. Park explorers have found arrowheads and pieces of hammered copper—evidence of these early peoples.

Commercial copper miners worked in the park area in the late 1800s and early 1900s but to little avail. Some of their diggings were located in the present-day picnic area at the center of the park, where the river used to flow. In 1902, miners diverted the river away from this area to prevent flooding of the mine shafts. They used dynamite to blast away a wing of basalt that had channeled the river through that area. This allowed the river to take the shortcut through which it flows today.

Recognizing the spectacular beauty of the park area, the state created Copper Falls State Park in 1929, setting aside 1,080 acres. The state has since more than doubled its area to more than 2,600 acres. The CCC did a great deal of work to make the park what it is today. Workers used local materials, including slate and sandstone, for building walkways, bridges, stairs, and the concession building.

TRAIL GUIDE
Doughboys' Nature Trail

Much of this 1.7-mile trail was built by the Doughboys—veterans of World War I—in 1920 and 1921, years before the area became a state park. The trail follows the rims of the gorge and crosses both rivers, giving the best views of the waterfalls and other features of the park. It is mostly level and easy but includes two steep ascents (or descents) into and out of the gorge, with a total change in elevation of about 120 feet.

The trail starts behind the concession stand near the central parking area, crossing the Bad River on a wide footbridge. Before crossing, look upstream and across the river at the high riverbank. It is made of reddish, silty sand and gravel deposited over thousands of years by the Wisconsin glacier.

Across the river the CCC trail to an observation tower splits off to the left, and the Doughboys' Trail continues to the right, east-northeast along the river. It is about a quarter mile to Copper Falls (Figure 2.11). At this point, the river has veered north to plunge over a ledge of basalt. In 1900 it was reported to be a 30-foot waterfall, and the western side (to your right) is still about that high. On the east side of the basalt block sitting near the center of the falls, the height has dropped to less than half of that measured in 1900.

How this drop occurred is an interesting story as told by retired but still very active Park Superintendent Kent Goeckermann. It begins in 1902 when miners diverted the river upstream, which changed the way it flowed downstream. It increased the velocity and force of the flow over the east side of the falls and lessened the force of the flow over the west side. The east-side flow drops over one ledge into a pool, where it hits a wall and veers north to make the second drop. For 85 years, the increased east-side flow undercut the wall, until in 1987 part of it gave way. Witnesses told Goeckermann that a slab the size of a boxcar

2.11 Copper Falls, Copper Falls State Park.

separated from the wall, tipped like a falling tree, and crashed onto the lower ledge of the falls, collapsing it by 12 to 18 inches. The slab shattered into pieces. A second, nearly identical rock fall occurred in 1989.

If it seems hard to imagine the riverbed changing so much in a relatively short period of time, consider that during spring flooding, the river tops the large basalt block that splits the falls. In fact, the Bad River was named for this sort of fury by early traders who found it impassable and had to look for other routes for transporting their furs. However, the carving of the falls and of the gorge does not wait for flooding, but is ongoing. During every winter, ice forms and expands in cracks and crevices in the basalt, weakening the rock, and every summer, flowing water continues its slow, steady dismantling of the rock. Says Goeckermann, "If you come to the canyon on a quiet day with no wind and little water flow, listen carefully and you will hear an almost constant fall of small rocks into the canyon and, not infrequently, you will hear large rocks tumble to the bottom."[11]

In the falls, the water has the color of root beer or copper, which is probably how the falls got its name (although some reports say the name came from the presence of copper deposits in the park). The Bad River originates in Caroline Lake, southeast of the park, and flows through many bogs. The water is stained by the tannins released to the river by decaying vegetation in the bogs.

As you continue downstream, keep looking to your right. The river has carved a gorge about 100 feet deep. Geologists think this is an area where one vast layer of rock slid over another as the Earth was tilted. This sliding action ground the rock at the fault into fragments that became breccia, which was easier for the river to carve.[12]

Nearly half a mile into your hike you come upon Brownstone Falls (Figure 2.12), where the Tyler Forks River meets the Bad River. This is a plunge of 30 feet over a basalt ledge. From the top of the cascade upstream from the falls, the river drops a total of 70 feet to the base of the falls. The river flows northwest across the ridge and valley structure described above, and this is what makes the series of falls.

Note that the rock on the left side of the falls is lighter-colored than the basalt of the Bad River gorge. This rock has been called red lava and probably came from a volcano that stood somewhere in this area a billion years ago. Geologists estimate that this red rock is more than 950 feet deep and likely came from a single eruption of the volcano.[13]

At this juncture, the Bad River turns 90 degrees to flow northwest across the grain of the land—the same direction as the flow of the Tyler Forks. The trail also turns to parallel the river. A short way down the trail from the falls is a vantage point where you can look down on Devil's Gate, a wall of sedimentary rock worn away by the river. This is one of the best views of rock layers that have been tipped steeply. Here the river encountered layers of conglomerate, shale, and sandstone, in that order, that were tilted to 86 degrees from horizontal.

About a quarter mile down the trail from the falls, you begin a gentle but long descent on slate steps into the gorge. These are the steps quarried and set down by the Doughboys more than 100 years ago. Not far from the base of these stairs is another sturdy footbridge across the Bad River, which offers another view of the Devil's Gate just upstream. Throughout summer and fall, the river

2.12 Brownstone Falls,
Copper Falls State Park

is usually placid at this point, but during spring flooding year after year, it has raged against this wall of sedimentary rock to create the gap you see.

Downstream from the bridge, the river meanders between steep banks of clay soil deposited by the glacial lake in the northern part of the park. From the footbridge, you climb out of the gorge on another set of slate steps. These stairs are steeper than those on the other side, but a bench provides a chance to rest and appreciate the ancient, beautiful river gorge.

At the top of the climb, the trail runs along the rim of the Bad River gorge where it merges with the North Country National Scenic Trail, which traces the northern landscapes from New York State to North Dakota. Continuing on the Doughboys' Trail about a half mile from this junction, you can take a spur trail to the right on a sturdy boardwalk across sensitive wetlands and terrain. It takes you through a spruce and cedar forest with a primeval feel to a viewpoint near the top of Brownstone Falls. This view of the falls is well worth the side trip, and you can also get a closer look at the reddish volcanic rock you saw viewing the falls from across the Bad River gorge earlier in the hike.

Back on the Doughboys' Trail, you pass the cascades above Brownstone Falls as you now walk upstream along the Tyler Forks River over the basalt layers built around 1 billion years ago by the Midcontinent Rift. Cross the river on a footbridge above the cascades and walk downstream for yet another view of the confluence of the rivers.

This overlook is just downstream of Brownstone Falls looking down the Bad River where it turns northwest (Figure 2.13). It is wheelchair accessible, as is this part of the trail, from a small parking area just to the south. It is arguably the most stirring view of all, in the heart of the park. To your right is the 90-foot, massive red lava wall next to Brownstone Falls. When the Bad River (flowing from your left) hit this wall, it turned left and burrowed its way through a lower ridge of softer lava rock, carving the notch through which it and the Tyler Forks together flow. The combined force of these rivers has created a boiling whirlpool just this side of the notch. Imagine the furious action around this whirlpool and notch during flooding. Beyond this point, the river slows to a more languid pace as it flows through the Devil's Gate and then meanders through the clay sediments to the north.

Just to the left of the Brownstone Falls you will see a large U-shaped gap in the red lava wall. This is an abandoned channel for the Tyler Forks River. There was once a rock barrier across the river at this point and the river eventually

2.13 The Tyler Forks River flows over Brownstone Falls (right) and meets the Bad River (left), where it veers northwest.

broke through it to find its present course dropping into the whirlpool. The Bad River would have been working on this wall from the other side as it swung northwest. Imagine that barrier still in place and the Tyler Forks gushing through the U-shaped gap to join the Bad River farther downstream. During flooding, the Tyler Forks River still gets high enough to send some of its water through that abandoned channel.

Another third of a mile closes the loop on the Doughboys' Trail as you arrive at the first bridge you crossed. About halfway on this stretch, you pass Copper Falls again, this time viewing it from behind the falls. This is the area where the rock wall collapsed, changing the shape of the falls, as described earlier.

One more side trip: Cross the bridge again and turn left to ascend the high ridge across the river from the concession building. This is a long climb up solid staircases. The trail at the top is named for the CCC workers who built it. On this trail to an observation tower, you are walking on the deep layer of reddish silty sand and gravel dropped here by the glacier beginning around 11,500 years ago. It took the glacier about 2,000 years to build it and since then, the Bad River has carved its valley into the mass of glacial till.

Straight Lake State Park

While Straight Lake State Park lies in the zone of the Midcontinent Rift, it was the far more recent Wisconsin glaciation that shaped this park. It is one of Wisconsin's finest examples of a glacially formed landscape.

When the glacier melted back, it had draped the land in a layer of till—sand, gravel, clay, and boulders. This new landscape, including the area around this park, was peppered with kettles—depressions left by slow-melting masses of remnant ice—of varying sizes. Straight Lake was formed in this way, as were almost all of Wisconsin's thousands of lakes. Since then, the lake's level has been raised slightly by the creation of a small dam at its eastern end, where it flows into the Straight River.

Straight River is literally quite straight and has flowed in its linear valley for at least 10,000 years. It was flowing long before the glacier finally shrank away to the northwest, for it originated as a stream flowing through a tunnel under the melting ice mass. The water was under high pressure from the weight of the ice and is thought to have gushed through its tunnel, carving a well-defined channel nearly 90 feet deep, up to half a mile wide, and 7.5 miles long. Over the centuries, the subglacial stream slowed and began to deposit the sand, clay, and gravel it was carrying, while overlying ice melted and dropped more debris onto the tunnel deposits. In this way, long ridges, called eskers, were built within the tunnel channel.

Hundreds of these tunnel channels were formed along the outermost edge of the latest advance of the glacier. Geologists think they carried huge volumes of water in sudden, short bursts.[14] Some of these flows carried large boulders to the margin of the glacier where they gushed out and then slowed, dropping the boulders and smaller debris in fan-shaped outwash deposits. The Straight River tunnel channel with eskers is considered one of the best-preserved postglacial landscapes in the world.

This is one of the least developed of all the state parks and one of the newest, established in 2005. It is a wilderness park with none of the amenities that state park visitors have become used to, beyond unpaved parking areas and modest directional signs. Within its borders is a segment of the Ice Age National Scenic Trail (IAT), which roughly tracks the farthest extent of the Wisconsin Glacier. It is the only trail that runs through the park.

TRAIL GUIDE

Straight Lake Segment of the Ice Age Trail

This trail is moderately difficult with several steep sections and rocks and roots to step over. The several bridges and boardwalks are well built but can be slippery when wet. The entire Straight Lake Segment of the IAT is 3.6 miles long, but I took just a part of that segment.

From the point where the road crosses the river near the parking area off Highway I, looking west-northwest, you get your first view of the tunnel channel (Figure 2.14). Around 10,000 years ago, you would be under the ice with a wide river rolling toward you. On the south side of Straight River, start the hike by passing through a gate in a fence. The first part of the trail is on private land used for grazing. Another fence and gate are about a third of a mile down the trail.

At the 0.75-mile mark, you cross onto state parkland. The well-worn trail, posted with yellow IAT markers, runs along the south side of the tunnel channel river valley. Near the one-mile mark, start looking for spectacular views of the tunnel channel to your right. Early spring and late fall when the leaves are down are the best times for this hike. To your left is hummocky land made up of hills, hollows, and little wetlands and ponds.

After 1.1 miles, the trail descends toward the valley floor, with more good views of the valley. After dropping, it climbs again to the top of a small esker and traces it for about a half mile, at which point it drops toward the river valley continuing westerly. About 2.5 miles from the start, the trail reaches the lake. On my hike I was lucky enough to see three trumpeter swans flying directly overhead going to the lake where they had nested on a small island. Their wings sounded like powerful fans and their calls were like bugles announcing their arrival.

At the lake, the trail crosses the small earthen dam on stepping-stones across the stream. The lake was dammed decades ago by loggers who needed a holding pond for their logs. When the state park was established, workers replaced the old dam and took great pains to make it unobtrusive, according to Dan Schuller, former director of the State Parks Division of the DNR. Indeed, you can hardly tell now that the dam exists.[15]

The trail runs along the north side of the pristine lake. It is undulating and mostly flat, passing through picturesque woods and a large plantation of red pines. The park is in a transition zone between prairie forest and northern hardwood forest and so contains species from both forest types. You will see old stands of oak on land that has been logged probably only once and long ago.

At the 3.3-mile mark, the trail veers away from the lake heading north into the woods after passing a beautiful wetland where the river comes into the west end of the lake. Look for boulders, some more than six feet in diameter, near the trail. They are glacial erratics, transported by the glacier and thought to have been picked up somewhere north of Lake Superior.

2.14 The Straight River valley is one of the best preserved tunnel channels in the world. A powerful stream under the glacier carved this valley.

Shortly beyond this point, the trail passes a parking area and crosses 280th Avenue. Within a quarter mile, the trail passes a number of impressive outcroppings of basalt, the bedrock laid down by multiple eruptions of lava from the Midcontinent Rift. The closest part of the rift was just to the northwest. Many of the outcroppings are covered with moss and other vegetation. A few even have trees growing from them.

Just under a mile from 280th Avenue, the trail reaches a small valley filled with boulders, some of them five feet in diameter. They are mostly basalt, fractured from the lava rock mass and moved here by the fast-moving subglacial stream that carved the tunnel channel. Among them are smaller reddish boulders of granite from deeper bedrock. Rhyolite boulders are also present here, providing evidence of explosive volcanic eruptions that punctuated the long period of slowly erupting basalt around 1 billion years ago.

The IAT continues northwest, or you can retrace your steps back to Highway I. For more views of the tunnel channel and eskers, turn east on 270th Avenue, just north of the Highway I parking area, and go about a half mile to the next section of the IAT, which runs south and southeast, continuing within the tunnel channel.

Interstate State Park

Interstate State Park is the official name for the Wisconsin side of a joint venture between Wisconsin and Minnesota, conceived in the late 1800s. Both sides of the park are usually referred to simply as Interstate Park.

It is fitting that Wisconsin's first state park, established in 1900, has some of the state's oldest geological features. The basalt formations that inspired the creation of the park were built starting around 1,100 million years ago during the Midcontinent Rift period. As each of the multiple lava flows from the rift cooled, the bodies of lava contracted and became segmented by long vertical cracks. Geologists refer to these cracks as joints, and they played an important role in the formation of the park's features.

Geologists think that when Cambrian seas covered most of Wisconsin, the highest points on the flooded land in the Interstate Park region, including Eagle Peak, were islands of basalt, occasionally battered by stormy seas. Where waves crashed against land, steep cliffs formed as boulders worn away by the wave action fell into the surf. Some boulders were broken into smaller pieces that were then tumbled against one another by the waves and, over thousands of years, became more rounded and smooth. These cliffside rock piles eventually were buried by sand and sediments as sea levels rose, and we can get cross-section views of these formations today within the park.

After the Cambrian sea departed, several more ancient seas advanced and retreated during the next 500 million years. They created more layers of sandstone and limestone on top of Interstate Park's bedrock layers of basalt. We have no record of this slow building process, due to the ensuing 400-million-year period of erosion. The next major act in the park's geologic story was the Ice Age.

The Superior lobe, the glacial lobe that covered the park area, was made up of sub-lobes, all moving in slightly different directions. In the park, visitors can find places where boulders embedded within the ice sheet scraped the rock over which the ice was flowing, leaving long scratches, called striations. While glacial movement in Wisconsin was predominantly southerly and westerly, the striations visible on Horizon Rock and in other areas of the park show that a sub-lobe moving through the park at some point in time was actually moving east-northeast.

At the peak of the Wisconsin glaciation around 20,000 to 24,000 years ago, the ice sheet over the park was somewhere between 1,000 and 2,000 feet deep. Beginning about 18,000 years ago, the ice began a slow retreat, then re-advanced several times and finally retreated completely from the state by 10,000 years ago. During that time, Glacial Lake Duluth occupied about half of what is now the Lake Superior basin to the north.

Between 16,000 and 10,000 years ago, Lake Duluth was dammed by the retreating glacier on its northeast side and, at its west end, was draining slowly through an outlet southwest of Duluth and down an ancient riverbed in northern Minnesota. The lake reached a level of 500 feet higher than present-day Lake Superior, so the Bois Brule River bed (now running north to the lake from the headwaters area of the St. Croix River, see Figure 2.2) was under water. Near that riverbed, the lake was held back by a fragile dam made of ice and newly formed moraine, about 80 miles northeast of Interstate Park. At some point, the lake's water breached this dam, sending a massive torrent of ice water southward over the highland where the Bois Brule River originates. Thus began the flood that carved today's St. Croix River valley and the gorge of Interstate Park (Figure 2.15).

2.15 The Dalles of St. Croix, Interstate Park.

When the great blast of water from the glacial lake arrived, it probably engulfed most or all of what is now the park, including its highest points. The huge volume of water and ice chunks coursed through the valley, tearing away much of the softer Cambrian sedimentary rock overlying the basalt bedrock. When the floodwaters reached the harder basalt, a different kind of erosion occurred. Rather than scouring off layers of rock, the ice water worked at the joints that had formed in the basalt when it first cooled. The water and ice chiseled at these joints until chunks of basalt came loose and were carried downstream by the flood.

The structure of the basalt rock made it easier for the water to carve vertically than horizontally, and as the channel became deeper, the water acquired more force per volume. Then the digging, chiseling effect became stronger and the carving quickened. In this way, the icy blast of water dug the deep, narrow gorge piece by piece. It probably took a few hundred years, which in geologic time was relatively quick.[16]

Another kind of carving that took place during the flooding created one of the park's most fascinating features: its famous potholes (Figure 2.16). In any flowing stream, you are likely to observe eddies, or places where the flowing water swirls, creating a vortex that draws water down in a spiraling fashion. Sand, gravel, and rocks can be pulled into this vortex and swirled against the streambed. During the postglacial St. Croix flood, gravel and boulders that got pulled into its eddies were whirled like natural drill bits, creating shallow circular depressions in the basalt bedrock that became deeper with sustained grinding. Over decades, the grinding action rounded many of these stones until they resembled cannonballs. They are called grindstones, and geologists have found them as large as 18 inches in diameter in Interstate Park.

Some of Interstate Park's potholes, easily viewed from the well-traveled Potholes Trail, are among the most impressive in the world, reaching diameters of 14 feet and depths of 18 feet. Even larger and deeper potholes were formed on the Minnesota side.

Still another feature created by the work of water and ice is the park's talus slopes, lying along the bases of many rock walls. Here fractured chunks of the basalt cliffs, called talus, were sent tumbling to the rock piles below by the ice-water flood thousands of years ago, and since then, by slower-acting frost, rain, moss, and other vegetation.

Within 1,000 years of the glacier's retreat, life returned to the Interstate Park area, eventually resembling the diversity of plants and animals that lived there just before the Ice Age. Near Lake of the Dalles, archaeologists have studied bison bones found in certain excavations along with arrowheads and tools used by Native Americans. The St. Croix River was part of the ancient route between Lake Superior and the Mississippi River. It is believed that the river was roughly the boundary between territories of the Sioux and the Ojibwe tribes at various periods, and thus it was also the scene of some battles between competing tribes.[17]

European immigrants brought lumbering, farming, and mining to the area. The St. Croix River again served as a major transportation route, this time for the removal of Wisconsin timber that was used to build thousands of houses and mansions from Superior to Chicago and from Milwaukee to Minneapolis.

2.16 One of many potholes in Interstate Park, drilled by glacial floodwaters thousands of years ago. At ground level, the pothole has a diameter of six feet.

What covers much of Interstate Park now is second- and third-growth forests that have their own remarkable beauty, if not the majesty of the ancient mature pine forest. Nevertheless, when you are hiking on one of the many trails in the park that pass through small groves of white or red pine, stop for a moment and close your eyes. Smell the scented air and listen for the delicate song of the wind in the pines. You might then imagine yourself to be in the vast, deep forest that once stood here atop gravel and soil dropped by the glacier, atop sandstone and limestone deposited by ancient seas, atop a bedrock of basalt laid down a billion years ago.

TRAIL GUIDE

Echo Canyon Trail

This loop trail is 0.7 mile long. It splits away from the Lake o' the Dalles Trail, which is accessible from the park's main parking and picnic area. Shortly after the junction, it crosses a partly moss-covered talus slope next to an ancient canyon wall. A short hike through the cool, moist air of the canyon takes you to the sandy riverbank where you get an excellent view of the basalt canyon wall across the river (Figure 2.17). Notice the multiple fractures in the wall and the pile of sharp-edged boulders near the riverbank. This gives you a sense of how the gorge was formed. Nowadays, rock climbers make use of the ancient joints and fractures.

From this point, the trail ascends back up onto the river bluff, merges with the Summit Rock Trail, and then curves back to close the loop at Lake o' the Dalles. The Summit Rock and River Bluff Trails are two other short loops that take you to splendid views of the St. Croix River gorge.

TRAIL GUIDE

Potholes Trail

This half-mile loop includes some high stone steps and uneven ground that make it a fairly rugged hike. It is also the western end of the approximately 1,200-mile Ice Age National Scenic Trail, which winds along the state's glacial moraine to its eastern terminus in Potawatomi State Park. The Potholes Trail takes you to the best views of the gorge with a deck and benches where you can sit and spend some time imagining the geologic drama that unfolded here.

Named for the park's extraordinary potholes, the trail traces a section of the rim of the gorge, at one point rising higher than 100 feet above the current river level. Potholes at this height tell us that the postglacial flood must have spread across and flowed over the entire park area for centuries, drilling potholes here and there, while the gorge was being excavated. As the gorge deepened, it took more of the water, and as Lake Duluth drained, the flood level dropped, leaving the potholes on the trail high and dry.

Also on this trail, look for tiny pits on the basalt surfaces, some of them filled with white, green, or pink minerals, some of them empty. These were formed around a billion years ago whenever a new layer of lava flowed over the land.

2.17 The canyon walls of Interstate Park, made of multiple layers of volcanic rock, are a popular destination for rock climbers.

Near the top of each layer of lava, gases bubbling up through the molten rock became trapped, and many were later filled with minerals such as white quartz and pink plagioclase feldspar. These mineral pockets in the basalt, called amygdules, are commonly visible in the park and indicate the boundaries between successive lava flows.

TRAIL GUIDE
Skyline Trail

This 1.6-mile trail runs from the Ice Age Visitor Center to a picnic and camping area. Most of it is a level trail along the southeast rim of the St. Croix River valley, with one steep section on uneven ground. It affords views of the vast river valley and three tributary valleys cut by streams that feed the river.

You reach the first of these after a half mile of hiking. At 0.7 mile, the Ravine Trail departs to the right, down into one of these tributary ravines, connecting with the beach area at Lake o' the Dalles, a small lake perched slightly higher than the St. Croix River and connected to it by Dalles Creek. When the leaves are down, this lake is visible from the trail. As you continue on the trail along the rim of the river valley, imagine the great flood of ice water that rolled through, filling the valley almost to the brim as the glacier retreated 10,000 years ago. The surrounding land would have been tundra over permafrost, hosting only extreme cold weather species such as lichens, some grasses, stunted alder and spruce trees, musk oxen, and caribou.

After 0.9 mile, the trail crosses another stream at the upper end of a second steep tributary valley. The water has worn away the layer of glacial till that covers most of the park, exposing a train of basalt rocks and boulders. The fact that they are rounded is evidence that this area was on the shore of at least one ancient sea where waves tumbled these rocks for centuries. After another 0.3 mile, the trail crosses a third tributary stream and valley on a boardwalk over an area soaked by spring water.

At about the 1.5-mile mark, the trail makes a right-angle turn to the right and heads down toward the river. Here it travels along a very deep fourth ravine. Across the ravine is further evidence of an ancient seashore—the remnant base of a steep cliff that stood in the waters of a Cambrian sea about 500 million years ago. The cliff was probably part of an island that included Eagle Peak, which is now buried just southeast of here.[18] Notice that the cliff wall is permeated from top to bottom with boulders and smaller rounded rocks. This was the mostly basalt rock pile that accumulated at the base of the cliff where the waves rolled and tumbled the smaller rocks, smoothing their sharper edges. Scan to the right and you can see layers of sandstone deposited by the Cambrian sea as it deepened. Sand from this deposit filled the spaces among the cliffside rocks and buried them, later forming conglomerate—a mix of sand, gravel, and boulders cemented by natural chemical processes. Much of it has now been eroded away, yielding this cross-sectional view of the ancient cliff.

The trail descends steeply for a tenth of a mile and then levels off as it passes some very large basalt boulders, some of which are erratics dropped by the glacier as it retreated to the northwest. It then crosses a pretty wetland that once carried meltwaters from the glacier, long after the great postglacial flood had subsided. At the 1.7-mile mark, the trail arrives at the East Pines picnic and parking area not far from the Pines Group Camp.

If you are up for a steep, rugged climb on a rocky trail, it is worth the effort to take the 0.8-mile spur trail to the top of Eagle Peak, the highest point in the park. The scent of pine is with you all the way as the trail curls among old white pines and ascends 120 feet to the rocky knob where you can get a clear 360-degree view of the park area. On some of the horizontal rock surfaces are faint parallel grooves—striations left by the glacial ice as it slid across the peak from the southwest.

3.1 Enee Point, a sandstone outcropping in Governor Dodge State Park

3

The Driftless Area

The Southwest Corner

When glaciers flow, as they have many times in Wisconsin, they change the land. They scrape away much of the land surface, toppling high points, filling valleys, and building ridges and hills with the materials they excavate. As they melt away, they leave broad deposits of sand, silt, gravel, and boulders. Years ago, this debris was commonly called drift.

From an astronaut's point of view you can see many of these glacial markings across Wisconsin: lakes, ridges, hills, and a lot of flat or gently rolling land (Figure 1.1). You can also see that the southwest fifth of the state looks different from the rest of Wisconsin. Most of that area has no drift or other signs of glaciation. It is the Driftless Area (Figure 3.2).

When the Silurian sea receded from Wisconsin more than 400 million years ago, the remains of a rich variety of life-forms had collected on the sea floor for millions of years. They were compacted to form layers of sedimentary rock averaging 600 feet thick across the state. This was mostly a hard rock called dolomite formed from limestone that had steeped in the magnesium-rich brine of the ancient sea over millions of years. The magnesium displaced some of the calcium in the limestone to convert it to dolomite.

Similarly, silica flowing in groundwater through buried sediments that are becoming limestone displaces some of the calcium to form another hard rock called chert, or flint. It appears as small or massive bodies within layers of limestone or dolomite. The Silurian dolomite that once covered the state was deposited on layers of older (Ordovician) sandstone, dolomite, and shale totaling more than 1,800 feet thick. In the southern part of the state, those layers were underlain by Cambrian sandstones laid down around 500 million years ago over a layer of igneous basement rock called rhyolite that had been spewed by violent volcanic eruptions more than 1,700 million years ago. Under all of these layers was 2,000-million-year-old granite bedrock.

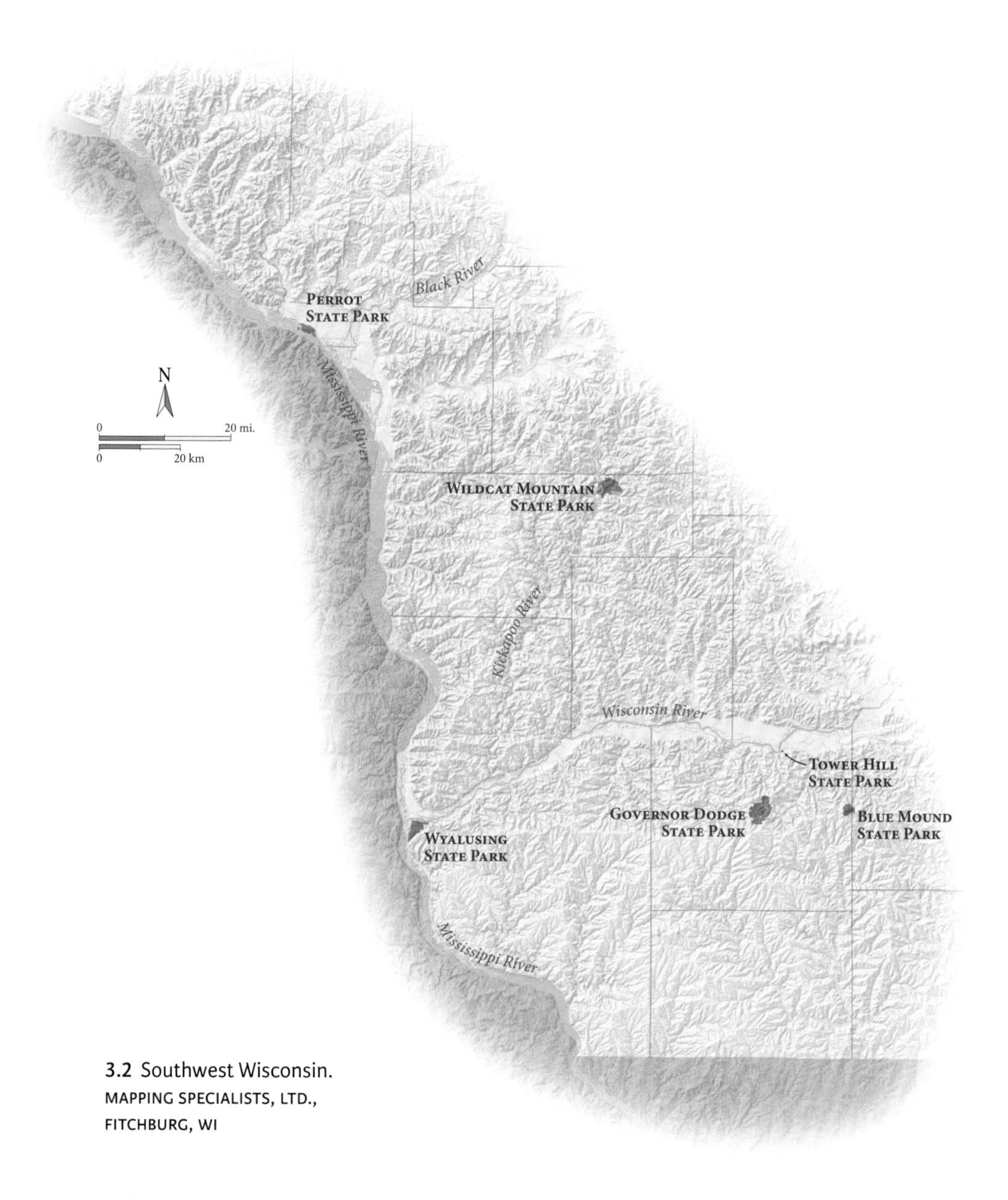

3.2 Southwest Wisconsin.
MAPPING SPECIALISTS, LTD., FITCHBURG, WI

With the departure of the Silurian sea, erosive forces slowly scoured away most of the Silurian dolomite layer, but its remnants are scattered across the state. Out of the layers of older rock underneath the Silurian blanket, wind, rain, and frost carved grand sculptures—steep promontories and dramatic outcroppings (Figure 3.1) set into a vast branching pattern of sharply defined ridges and valleys like those that now make up the Driftless Area. For millions of years, most of Wisconsin probably resembled a less-eroded version of today's Driftless Area. Just before the Quaternary period, some 2 million years ago, the climate was mild and moist. In southern Wisconsin, the highlands were covered by prairie and oak savanna, and the deepening river valleys were forested with trees similar to modern species.[1]

Then came the glaciers. They plowed across the state to the north and east of the Driftless Area but time and again stayed clear of it. The question of why all the glaciers avoided the Driftless Area has fascinated geologists, and part of the answer involves geographic luck. As the ice advanced, it was channeled by lowlands, and with each successive advance, the lowlands were carved deeper, making stronger channels for the next advance. Thus north of the Driftless, the Lake Superior lowland steered the ice sheets primarily westward. East of the Driftless, the lowland now occupied by Green Bay channeled the ice generally southward.

Another factor was glacial stamina, or lack of it. Had any of the glaciers had more time and more volume, they might have overwhelmed the northern highlands and spilled southward across the Driftless. One or more glaciers did that to a limited extent. In the case of the Wisconsin glaciation, there was not enough mass behind the southern margin of the Lake Superior lobe to make that happen. Before the ice mass could grow more, the climate warmed and it began to melt back. You could say the glacier ran out of steam, but it would be more fitting to say it ran out of ice.

Some popular literature claims that the Driftless Area was like an island surrounded by ice at various times. However, based on their ability to determine the ages of different rocks, geologists think this was never the case. Glacial till to the south of the area is from a different time period than the till on other sides of the Driftless Area, so these till deposits could not all have been laid down at the same time.

For 2 million years, the Driftless Area avoided glaciation, but it did not avoid change. While the ice mass was plowing and carving the rest of Wisconsin, water continued to sculpt the Driftless, and so did ice, if not in a massive form. In the frigid climate of glacial times, the water frozen in soil and rock crevices did its part in breaking down rock and changing the landscape. In the crevices, water froze and expanded year after year for centuries, gradually fracturing and prying rock segments apart. The permafrost, or frozen soil, was hundreds of feet deep and during summer only a few feet of it could thaw. With enough rain, this thawed layer would become saturated and, on steep slopes, could give way in landslides. In these ways, the glacier indirectly changed the face of the Driftless Area.

As the glacier melted back, it continued to leave its mark on the area. The valley of the lower Wisconsin River is now a broad, flat passageway that bisects the Driftless—an uncharacteristic feature of the landscape (Figure 1.1). This is because it was a conduit for massive volumes of water and ice chunks flowing away from the melting glaciers of the Pleistocene, especially the last one, beginning about 12,000 years ago. Torrents of postglacial ice water bored out the valley, carving it hundreds of feet deeper than it is now. Over the centuries since the most recent glaciers, the river steadily hauled huge volumes of gravel, sand, and silt to fill and broaden the riverbed.

The Upper Mississippi River was also carved by glacial meltwaters. Its present-day course south of the Wisconsin-Illinois border is thought to have been established during the first of the four major glaciations, beginning about 2.4 million years ago. Evidence indicates that before then, it originated in north-central Iowa and flowed southeast to its present course on the Illinois border. That earliest glacier covered the ancient valley and pushed its way to what is now the Wisconsin-Minnesota border. The meltwater flowing along the eastern edge of that glacier might have formed the modern Mississippi Valley.

However, intriguing research conducted since 2010 offers evidence that what is now the far Upper Mississippi, instead of flowing south all the way to Illinois and beyond, once veered east at the point where the Wisconsin River now joins the Mississippi. According to this explanation, long before the most recent glacier, the Wisconsin River did not exist as we know it now. Instead, the ancient Mississippi turned left and flowed east across the state in the bed of the present-day lower Wisconsin River. Researchers led by Eric Carson of the Wisconsin Geological and Natural History Survey compiled data, based primarily on extensive core drilling, that indicate that the bedrock under the sand and gravel riverbed of the lower Wisconsin dips gently to the east.[2] This and other factors (see the Wyalusing State Park story, page 99) convinced the researchers that a river must have flowed east within the lower Wisconsin River valley until sometime between 2.4 million and 760,000 years ago. They call this hypothetical ancient stream the Wyalusing River.

These findings also imply that a high ridge of resistant rock extending from today's Military Ridge in southern Wisconsin once continued west across today's Mississippi Valley, forming the barrier that caused the ancient Wyalusing River to turn sharply to the east. There was likely a stream draining the south side of this highland through the present-day Mississippi Valley on the Illinois-Iowa border. The headwaters of that stream would have worked their way upstream via slow erosion, while the Wyalusing River was wearing away the ridge that caused its sharp left turn. Together, these erosive processes eventually broke through the resistant ridge, and the water from the Wyalusing began flowing south instead of east. Geologists colorfully call this process stream piracy.

Thus the Wyalusing abandoned, or was pirated away from, its eastern stretch. But while the lower Wisconsin River's bedrock still dips to the east, the riverbed—made up of thick deposits of sand and gravel on top of the bedrock—now dips gently to the west. This is probably because, over thousands of years, glacial lake deposits and outwash from glaciers retreating eastward piled up across southeastern Wisconsin. Eventually water started draining off this glacial debris to the west, through the lowland created by the ancient Wyalusing River. The new stream thus modified the old riverbed and became today's Lower Wisconsin River.

At the end of the glacial period, the Mississippi probably carried many times its current volume of water. One reason is that it was helping to drain Glacial Lake Agassiz, an enormous body of meltwater that once covered all of northwestern Minnesota, eastern North Dakota, and large areas of Ontario, Manitoba, and Saskatchewan. This and other sources of meltwater broadened and deepened the Mississippi Valley starting about 12,400 years ago. At that time, a set of braided streams, or streams with many interwoven channels, flowed through the valley. During torrential flows, such as when an ice dam would burst, water carved away at the sandstone bluffs on the valley sides for decades or centuries, widening the flow. Over time, the ice melted back and the main flow narrowed, carving deeper channels in the valley floor and leaving broad, flat terraces along the side of the Mississippi and other large rivers, including the Wisconsin.

During the melting, streams and rivers were flowing full force much of the time. Whenever flows subsided for any period of time, vast expanses of silt deposited on floodplains would be left to dry, especially along the Mississippi River and its tributaries. The strong winds blowing at the time would then stir up massive dust clouds and drive them eastward. These windblown sediments accumulated on the barren highlands east of the rivers. The result was thick deposits of rich silty soil called loess, as deep as 60 feet in some areas closest to the ancient floodplains and thinning to inches deep in central Wisconsin. Generations of Native American and European American farmers of the loess-covered lands have used this gift of the glaciers—the fertile topsoil that covers much of Wisconsin.

As the postglacial climate gradually warmed, tundra across the state gave way to boreal forests, which in turn were eventually replaced by diverse coniferous and deciduous forests in the north, and prairie and oak savanna in the south. In the Driftless Area, plant and animal communities probably survived the glacial times, although there has been some disagreement among scientists on this matter. Some argue that the ice-free area of Wisconsin was warm enough during the moist summers to support vegetation similar to what is found there today.[3] Others contend that plant cover would have been limited to hardy ground plants of the tundra and scatterings of spruce stands in the valley bottoms or in other sheltered areas.[4]

In the postglacial Driftless, animal communities gradually grew more diverse, eventually including deer, foxes, bobcats, opossum, beavers, squirrels, rabbits, chipmunks, voles, shrews, tree frogs, red-bellied snakes, and many other species. Birds included turkey vultures that favored bluffs and crags for hunting and still do today, as well as grouse, turkeys, killdeer, mourning doves, woodpeckers, and tree swallows. In the deep, cool valleys where water seeps through sandstone cliffs, rare ground plants took hold thousands of years ago, and a few of those ancient species remain today, including Sullivant's cool-wort (*Sullivantia sullivantii*) and the threatened species muskroot, or moschatel (*Adoxa moschatellina*).

As the glacier retreated from lands around the Driftless Area, hunters of the Paleo-Indian tradition entered Wisconsin. It is possible they were first attracted to the area because it was not laid bare of vegetation as were lands around it, and was thus more hospitable for the animals they were hunting. The hunters were following the wooly mammoths, musk oxen, mastodons, and other big-game animals that were following the glacier. In the Upper Kickapoo River valley, in the heart of the Driftless, archaeologists have registered more than 450 sites of early occupation dating between 850 and 12,000 years ago. The sites have provided a wealth of information about the area's early Native Americans, from the Paleo-Indians through the Woodland peoples. They include rock shelters, camp and village sites, burial mounds, and petroglyphs—images etched into rock. Similar sites can be found in the parks of the Driftless Area.

The Native American tribes living in the area when Europeans arrived were the Sauk, Meskwaki (Fox), and Ho-Chunk. In addition to hunting and farming, these three tribes all took part in the mining of lead in the far southwest corner of Wisconsin. Later, they were joined by European American miners who arrived in the lead rush of the 1820s.

Lead and zinc were products of a much earlier time, forming around 245 million years ago when hot, briny water solutions containing lead, zinc, and other minerals flowed into the region from 600 miles south of Wisconsin. The upheaval of land in that area (now Tennessee and Arkansas) forced groundwater down to deep rock layers where it was heated. The searing-hot water dissolved minerals from the rock, and gravity and pressure pushed the mineral solutions northward under Illinois and upward to the limestone and dolomite layers under the Driftless. There the solutions cooled in crevices and caves and the minerals precipitated to form solid deposits of lead and zinc.

For at least 8,000 years before the lead rush, native tribes had been mining lead in small quantities. They used it to make body paint and decorate clothing, and they buried the finest pieces with their dead.[5] When French explorers arrived in the 1700s, they taught natives how to smelt lead, and Ho-Chunk, Sauk, and Meskwaki people began mining lead on a larger scale. They learned to make their own musket balls, which gave them something other than furs to trade and saved them from having to buy the musket balls used for hunting and tribal defense. It was the women who did the mining. This is thought to have been an extension of their having dominion over tribal farm fields and maple sugar production.[6]

The European exploitation of lead began in 1820 and peaked in the 1840s. As lead mining was fading, zinc deposits were discovered and another boom ensued. Zinc mining peaked in 1915, and all lead and zinc deposits were economically depleted by the 1970s.

As French fur traders were expanding their territories, the Ho-Chunk were displacing the Sauk and Meskwaki tribes that had lived in the Driftless Area for centuries. The Ho-Chunk were in turn displaced by Europeans in the 1800s, forced to move west of the Mississippi in 1837 but returning in later decades. In the 1840s, the lumbering companies came to clear the forests. Soon after, European immigrants came to farm the land.

The present-day Driftless Area is dressed differently than it was in pre-European settlement days, now mostly covered more by farms and forests. Before settlement, it was largely given to prairie and oak savanna with forests growing in the river valleys. Seasonal fires were a part of that ecosystem, just as they were on the vast prairies surrounding the area, and for thousands of years, fire kept forests from growing on the flat highlands. European settlers suppressed enough of the fires to take them out of the ecological formula, and thus the settlers and their descendants helped to convert some of the upland prairies to forests.

Under the field and forest cover, however, the Driftless looks much the same as it did a few thousand years after the last glacier. It is a largely flat upland on Paleozoic bedrock with an intricate network of valleys cut deeply by streams and rivers over millions of years. Because of this extensive drainage by streams, no natural lakes and few large wetlands remain in the area. The landscape includes high bluffs, rocky ledges and outcroppings, buttes, spires, and isolated mounds. Perhaps it was the early French explorers who brought the moniker "Coulee Country" into use. *Coulee* comes from the French word *couler*, which means "to flow" and probably referred to the streams trickling through the deep valleys. However, *coulee* is now a common term for the narrow, steep-sided valleys themselves.

The Driftless is one of Wisconsin's best places to travel through time. You can venture back 500 million years or more by going into the deepest valleys. As you climb out of a valley, you return through the eons to the present, and you can go right back again by descending into the next coulee. The state parks of the Driftless Area give you plenty of opportunities to do just that.

Perrot State Park

Approaching Perrot State Park from the east on State Highway 35, you cross an eight-mile stretch of flat land, leaving behind a range of hills north of La Crosse that are classic Driftless country. At the end of that flat stretch is the town of Trempealeau on the Mississippi River, and looming just west of the town are the nine named peaks of Perrot State Park, reaching from 350 to 520 feet above the river. They make up a remnant cluster of Driftless peaks standing above a crescent-shaped plain that is nestled into the Driftless Area to the north and east.

Perrot State Park is located in the Driftless Area, so was never touched by glaciers. However, the waters flowing away from the ice mass played a major role in the park's formation. The entrance to the park, like the village of Trempealeau, sits on a terrace 60 feet above the Mississippi River. Around 400 million years ago, it would have been buried under more than 1,000 feet of rock—layers of sandstone capped by a thick mass of dolomite, most of which was later carried away by erosion. The mounds in this park are mostly Cambrian sandstone and have survived erosion because they are capped with resistant Prairie du Chien dolomite.

The first of the Pleistocene glaciers that inched across the land starting about 2 million years ago is thought to have pushed from the west to the general area of the present-day Mississippi River valley. As that first glacier began to melt, a river formed at its eastern edge and massive amounts of water from the glacier carved the valley there. At the time, the peaks and ridges of Perrot State Park were not isolated as they are today, but were part of the Minnesota bluff area that now rises across the river. The river, meanwhile, flowed through the crescent-shaped flat area north and east of the park bluffs (over which you drive to get to Trempealeau). Tributaries to the river began carving channels that separated those bluffs from the Minnesota bluffs, and these channels became wider and deeper over thousands of years.

With the retreat of one of the earlier glaciers, the Mississippi was carrying loads of sand, silt, and gravel from the melting glacier. At the north end of the crescent-shaped area, the Trempealeau River flowed down out of the highland into the Mississippi. The large flow of glacial debris, along with sediments from the Trempealeau River, gradually built a low dam across the crescent area, causing the Mississippi to veer southward into those channels carved by tributaries (and onto the river's present course) about 50,000 years ago. This left the Perrot peaks stranded on flat land south of the rest of Wisconsin's Driftless Area and separated from their original body of rock by the Mississippi River.

One of the most visited peaks in the park is Brady's Bluff. At the foot of that bluff on its west side is Trempealeau Bay, part of a vast bottomland formed by the confluence of the Trempealeau and Mississippi Rivers. Across the bay is

Trempealeau Mountain, the largest island in the Mississippi (Figure 3.3). It rises 388 feet above the river and is protected as a state natural area. Thus there are no designated park trails on the island and hiking there is not allowed.

On one of my trips to the park, I was fortunate to run into park naturalist John Carrier, who hiked with me to the summit of Brady's Bluff. Looking down at expansive Trempealeau Bay, he told me that part of it was created by a levy built across the bay by the US Fish and Wildlife Service to impound water that serves as a nesting and staging area for waterfowl.

"It's an important refuge," Carrier noted. "In the spring, that water will turn white with migrating trumpeter and tundra swans and American white pelicans."[7] Several thousand of these and other birds have been counted in the bay during migration.

Carrier's work as a naturalist has included the exploration and study of Trempealeau Mountain, which is not open to hikers unless they are accompanied by a park official. He and fellow researchers have examined fire-charred rock under sediments on the island—evidence of ancient campfires used by very early Native American travelers.

If you hike to Brady's Bluff, take time to view the mighty Mississippi and imagine how it has changed since it was a broad, braided outwash stream flowing away from the retreating glacier. Whenever a glacial impoundment such as the immense Lake Agassiz drained through the valley, huge torrents of ice water overwhelmed the shallow braided channels and bored through the valley, widening the riverbed. Such floods crashed against the sides of the bluffs, which were made of easily eroded sandstone lying under the more resistant layer of dolomite on the bluff tops. When the water would cut deeply enough into a bluff side, the overlying layers under their own weight would break away from the bluff and plunge into the river. In this way, the dramatic vertical bluffs we see now in the Upper Mississippi Valley were formed.

When the volume of water dropped and the flow narrowed, it would cut the riverbed deeper through its main channels. Soon after the Wisconsin glaciation, roughly 10,000 years ago, the Upper Mississippi was about 500 feet deeper than it is now, which means the bluffs of Perrot State Park were close to 1,000 feet above the water.[8] Since then, sediments from upstream and from tributaries have filled the valley to its present level.

In the park area, the earliest arrival of humans is estimated to have been 7,000 years ago, during the Archaic period. These early travelers hunted

3.3 Trempealeau Mountain, one of several peaks in Perrot State Park.

mostly bison, elk, and deer, gathered nuts and berries from the forest, and fished the rivers. The river environment has provided a rich variety of resources for native peoples since that time. On Trempealeau Mountain and across the rest of the park are many burial mounds built by Woodland cultures beginning about 2,500 years ago and later by the Hopewell culture between 100 and 400 CE. Trempealeau Mountain has been sacred ground for the native Ho-Chunk people.

The river provided a major trade route between north and south. The variety of pottery, pipes, ornaments, and other items found in the park by archaeologists indicates that the trade network was widespread. Many of the goods came from as far away as Wyoming and Appalachia. People of the Oneota tradition lived in the area between 1000 and 1500 CE. They lived in semipermanent villages on the riverbanks. While they continued to hunt, fish, and gather forest resources, they also began farming, growing corn, beans, squash, tobacco, and sunflowers. The park's nature center has an excellent display with more details about the area's Native American history and culture, as well as about archaeological work in the area.

One of the first European explorers of the Upper Mississippi Valley was Nicholas Perrot, a French trapper. He joined in the native fur trade and negotiated treaties with some of the resident tribes. In the winter of 1685, Perrot and his group made a winter camp in the area near the historical marker near the park's entrance. Fifty years later, French forces built a fort on the site.

In the late 1800s, John A. Latch, a wealthy businessman from nearby Winona, Minnesota, dedicated himself to acquiring land on the banks of his beloved Mississippi River. His sole purpose was to donate the land to the states so that it could be enjoyed by the public. Of the 18,000 acres he donated, 1,010 became Perrot State Park in 1918. Latch's wish was to have the new park named after the famed French explorer and trader.

As in other parks, the Civilian Conservation Corps (CCC) made its mark on Perrot State Park, building a camp there in 1935. The CCC first transplanted trees to save them from flooding when the Trempealeau lock and dam were built on the river. They also quarried rock from area bluffs and used it to build steps on the park's most traveled trails, as well as the shelter on top of Brady's Bluff.

TRAIL GUIDE
Brady's Bluff and Riverview Trails

Three trailheads access Brady's Bluff. On the east and west trails up the bluff, you pass through the Brady's Bluff State Natural Area, which protects a dry prairie on a steep, southwest-facing Mississippi River bluff. The prairie contains more than 100 species of native Wisconsin plants and several rare animal species, including butterflies and the threatened wing snaggletooth land snail. The trail up the north, or back, side of the bluff is not as steep and rugged as the east and west trails, nor is it as scenic, being entirely wooded.

I took the East Brady's Bluff Trail, starting from the parking area near the park headquarters. Brady Bluff is 520 feet high. This is a difficult trail because of the steep sections and narrow, rocky stretches. However, it offers a superb sampling of the park's features, from deep streambeds to a high bluff-top prairie.

Shortly beyond the east trailhead is a junction with the Perrot Ridge Trail. Bear left to stay on the Brady's Bluff Trail, which shortly thereafter crosses a sturdy footbridge across a small stream and climbs one side of the stream valley. For about a half mile, the trail switchbacks its way up the bluff, passing water-worn sandstone outcroppings that show the river was once flowing at this level during the retreat of the glacier. It then levels off somewhat for a spectacular view of the Mississippi River and the Minnesota bluffs.

From there, the trail gets steep again as it climbs through a goat prairie toward the rocky bluff (Figure 3.4). At the top end of this stretch are steps going to the summit. All the way, you are walking on dark-colored soil called loess, made of the silt once dropped on the river bottom by glacial meltwaters. As the river dropped, loess-covered floodplains dried out and massive clouds of the silty material were blown by high winds across the postglacial landscape where they accumulated in layers up to 60 feet thick just east of the river.

After three-quarters of a mile, you reach the summit and the shelter built in the 1930s by CCC workers. This vantage point offers a panoramic view of the Mississippi Valley to the south and Trempealeau Mountain and Bay to the west (Figure 3.3).

3.4 A bluff-top dry prairie, or goat prairie, in Perrot State Park.

The West Brady's Trail offers an experience much different from the east trail. On this descent, you quickly cross the eons. At the top of the bluff, you are standing on Prairie du Chien dolomite deposited by Ordovician seas more than 400 million years ago. Descending, you pass through layers of sandstone, older dolomite, and still older Cambrian sandstone laid down as long as 600 million years ago. Each foot of layered rock represents as much as 300,000 years of geologic time.

The trail includes several sets of steep stone staircases, built by CCC workers, and newer wooden stairs built by the Wisconsin Conservation Corps. This trail too offers spectacular views of the river valley. All along the trail are gnarly old cedars, seemingly growing right out of the rock. Many are probably hundreds of years old. As on the east bluff trail, this one passes exquisite sandstone outcroppings carved by flowing water centuries ago.

About three-quarters of a mile from the bluff top, the trail levels off somewhat and winds gently down through a hardwood forest. If you time an autumn visit right, you will encounter blazing fall colors on this trail. All along this west

3.5 Canyon on the west side of Brady's Bluff, Perrot State Park.

bluff trail, you get vistas of sheer cliffs where the sandstone is gradually eroding away in chunks, with fresh sandstone faces showing here and there. It makes you aware that these bluffs are continuing to erode and will be part of the river floodplain someday.

About one mile from the bluff top, the trail has descended into a beautiful little canyon with a primeval feel to it (Figure 3.5). Here the bluff is being eroded by moss and other vegetation as well as by rain, snow, and frost. Shortly beyond this point, the trail exits the woods to a boat landing and parking area. From here, you can take the Riverview Trail back to the park office area.

Wildcat Mountain State Park

From the top of Mount Pisgah, a sandstone mound in the southwest corner of Wildcat Mountain State Park, you can get a sense of what Wisconsin was like before the glaciers gave the land a makeover (Figure 3.6). A distant ridge with forested hollows etched into its flank rises abruptly from a valley floor where the Kickapoo River meanders. All is quiet but for the rustling of leaves in the breeze, and you might see a vulture soaring near, scoping rocky outcroppings on the mound for something to eat.

Wildcat Mountain State Park is in the heart of the Driftless Area, situated on a long ridge that borders the valley of the Kickapoo River. Wildcat Mountain, Mount Pisgah, Little Pisgah—all the named peaks in the park—are part of this ridge system running southwest to northeast.

Not far north of the park are the headwaters of the Kickapoo, named by Algonquin-speaking Native Americans for "he moves about, standing now here, now there." It is an apt description. The river winds crazily through its valley and nearly meets itself coming in several places. Within its 65-mile-long drainage area, the river flows 125 miles. Feeding it are dozens of tributaries flowing from the coulees on either side of the valley. The pattern of this river system is called dendritic, resembling the branches of a tree coming off a major limb, and this too is typical of the Driftless Area.

The placid nature of the river belies the fact that it represents an enormous demolition project—the dismantling and hauling away of vast volumes of dolomite and sandstone that once made up the mantle covering this part of Wisconsin. As you stand near the Kickapoo, or if you take to it in a canoe as many

3.6 The Kickapoo River Valley lies adjacent to the bluffs in Wildcat Mountain State Park.

do, look up at the bluffs on either side of the valley and realize that all the land in the area was once at that level and higher. It took around 400 million years for the streams to carve an immense arterial system into the ancient bedrock.

On the valley floor, you are among Cambrian rocks, made of grains of sand carried to an ocean shore more than 500 million years ago. As you climb toward a ridgetop, you move forward in time and on the highest peaks you are standing on Prairie du Chien dolomite, laid down over millions of years until about 480 million years ago by an Ordovician sea. Since then, additional layers of sandstone, dolomite, and shale were deposited by later seas and carried away by erosion. While the parkland seems to be undisturbed, much has happened here.

The first signs of humans in the park area were left by Woodland Indians who hunted in the region as long ago as 2,500 years. At that time, they had established settlements in the Tomah area northeast of the park and to the southwest where the Kickapoo joins the Wisconsin River. Archaeologists have found no evidence of such settled camps in the park area, leading them to think it served as a seasonal hunting ground for people passing between the more permanent settlements.

In the 1800s, after the Ho-Chunk ceded the area, loggers moved in to clear much of the forested areas of the park. They floated their cut logs down the Kickapoo to the Wisconsin and on to mill cities in the southwest. European immigrants moved in soon after to farm the land.

As with other Wisconsin state parks, this one got its start when a landowner who had fallen in love with the area's natural beauty donated the land to preserve it. In 1938, Amos Theodore Saunders gave 20 acres in the park's lower picnic area to the state. Ten years later, the state established a state park on that land along with 60 acres donated by Vernon County. The park has since grown more than fourfold, now covering more than 3,600 acres.

The park is ironically named after a bobcat that was killed sometime in the 1800s by local farmers displeased with the cat's choice of their sheep for its prey. The hunting party found the cat on the ridge near the present-day park's main campground, and thereafter called it Wildcat Hill, which later became Wildcat Mountain.

TRAIL GUIDE
Hemlock Nature Trail

This 1.3-mile loop trail to the top of Mount Pisgah is considered rugged and difficult with a total elevation change of 365 feet, but if you take your time, it is manageable. It begins at the original site of the park—the lower picnic area on the Kickapoo River. The first stretch is a flat, grassy, wide trail along the river, and it makes a good, short, easy hike if you do not want to climb. As it begins to ascend Mount Pisgah, the trail becomes more narrow and steep, beginning with a set of steps.

Near the top of the steps, the trail splits. I took the right branch, which switchbacks up the bluff. On the first switchback, it circles back through a beautiful stand of white pines to a place overlooking the river. These pines along with the virgin hemlock along the trail are part of a rare old-growth forest stand, as this part of the park was never logged. The sandstone underfoot is porous and holds water, which helps to keep the slopes cool and moist. Because this area is rare and pristine, it has been protected as a state natural area.

The trail meanders away from and back toward the river several times as it crosses the little valleys cut into the land by seasonal streams that flow into the Kickapoo. On this trail, you feel that you are climbing through time—many thousands of years for every foot of elevation—as you rise among the layers of Cambrian sandstone.

The state natural area through which the trail runs hosts a number of rare plant species, some of which were around before the glaciers, including Sullivant's cool-wort (*Sullivantia sullivantii*) and moschatel (*Adoxa moschatellina*). Both are low-growing flowering plants that like the cool, moist conditions found in the area now as in the distant past. Look for walking fern (*Asplenium rhizophyllum*), whose long arching leaves radiate from its roots. Where the leaf tips touch the ground, they sprout new plants. It can be found "walking" across rock outcroppings.

3.7 This sandstone outcropping was eroded by wind, water, and frost.

As the trail tops the ridge, it becomes a gentle, nearly flat hike of about 100 yards to a lookout with railings on the edge of the bluff facing north. This vantage point stands at 1,225 feet above sea level, some 365 feet above the river valley, allowing a remarkable view of much of the rest of the park and the Upper Kickapoo Valley (Figure 3.6).

From here, the trail switchbacks down the other side of the bluff. Notice the acorns crunching under foot. White oak is a prominent tree species in the park. The trail passes some spectacular sandstone outcroppings, intricately eroded by wind, water, and ice (Figure 3.7). Spend some time here to find an array of rock layers, crevices, and cavities that show what wind and water can do to rock over millions of years.

This stretch of the trail is shorter than the path that took you to the top of the bluff. It angles north to complete the loop near the steps that take you back down to the river level and the walk back to the picnic area.

The Old Settler's Trail is not as pristine as the Hemlock Nature Trail, but it is still an interesting hike. It is a longer trail, a 2.5-mile loop, and is a bit more rugged with a total elevation change of 390 feet. It starts at the top of the bluff on the northwest end of Wildcat Mountain and takes you down into a hollow typical of the Driftless Area. It has more steep stretches than the other trail, winding among ridges and hollows. The farthest point out on the trail is Taylor Hollow Overlook, which gives you another view of the Kickapoo Valley and the Village of Ontario.

Tower Hill State Park

When taking a trip, it usually makes sense to set out with a map and a plan, but sometimes you pass an interesting hill, ridge, or ravine and just want to park the car and explore. Tower Hill State Park is a good place to satisfy such an urge. The hills that beckon as you drive east across the Driftless Area on State Highway 14 are those that rise abruptly on the south side of the broad, flat Wisconsin River valley.

The Lower Wisconsin River's vast plain was built as the last glacier retreated some 10,000 years ago. For hundreds of years, the flood of ice water first excavated a riverbed about 200 feet deeper than what you see today, then filled and broadened it as the meltwater flow slackened. As the wall of ice retreated north, the river narrowed and carried less outwash and instead of building up the floodplain, began eroding it again. It thus carved a new, narrower riverbed in the outwash plain it had created. This left level terraces, or benches, on either side of the river where it had once flowed more broadly.

Native Americans and European settlers found these river terraces to be good places to camp and to build settlements. Much later, railroad and highway builders found the terraces useful, and today, towns such as Arena and Spring Green are located on the terraces of the Wisconsin River. Between the two towns, you can take County Highway C from Highway 14, winding your way into the highlands. It is just two miles from the junction to Tower Hill State Park.

The park is a historic site, famous for the tower built in 1831 and 1832 for the purpose of making lead shot of all sizes. This involved the painstaking job of digging a 120-foot vertical shaft and a 90-foot horizontal tunnel—work completed by just two men over 187 days. They used hand tools only, worked when the weather permitted, and took time off to fight in the brief Black Hawk War of 1832.

The park commemorates the tunnel and tower builders, the process of making the shot, and the history of the business. It is a fascinating story, but it's not all the park has to offer. You can also get sweeping views of the vast river valley from a trail along the bluff top, 150 feet up from the river. You can hike along the edge of the wild and pristine bottomland of the Lower Wisconsin River, hearing and seeing birds of dozens of species, along with other wildlife. Between those trails, you can clamber up or down the bluff on stairs, traveling back and forth in time as you cross layers of sandstone representing many thousands of years for every foot of elevation change.

This is one of the smaller state parks with a set of unnamed trails that can easily be hiked within a morning or afternoon. All but one loop of the trail system are covered in the following Trail Guide.

TRAIL GUIDE
Shot Tower Trail

The most direct trail to the shot tower begins at the park pavilion. Officially unnamed, it is a steep paved trail going more or less straight up the hill, just 0.2 mile to the tower. The last part of the ascent is a set of steps that takes you to a building attached to the 60-foot wooden tower (this one a facsimile of the original) that sits over the 120-foot shaft used for shot making. Just before you reach those steps, you can take a short side trail to the left. It hooks around the summit of the hill and then runs for a few yards along the face of the bluff to the base of the wooden tower. This side trail hugs a sandstone overhang that vividly reveals the beds of sandstone laid down around 500 million years ago as an Ordovician sea advanced from the south (Figure 3.8).

Back on the main trail, after spending time at the shot tower display, you can follow the trail along the top of the bluff, bearing left at intersections, and within a half mile, the trail curves gently down to the river bottom. It lies on an old oxcart road that continues along the base of the bluff—a wide, grassy trail that makes for a pleasant hike. I was there in early spring and heard a cacophony of birdsong, including geese, ducks, sandhill cranes, herons, turkeys, woodpeckers, and robins and other songbirds.

This trail runs for a third of a mile to the tunnel that was dug into the bluff to the point where the vertical shaft descended. Shot makers released drops and dollops of molten lead from the top of the tower, and as they plummeted through the shaft, they formed into spheres. At the foot of the shaft, they splashed into a pool of cold water and solidified into balls of varying sizes to be used as lead shot. The end product was retrieved through the tunnel and shipped downriver or hauled back to the top on the oxcart road. The tunnel is now blocked at its inward end, but you can still walk about 30 yards in under the bluff.

Returning from the tunnel on the old oxcart road trail, a little over a hundred yards from the tunnel is a set of steps that switchback up the face of the bluff. These are steep, so be careful and take your time. You are heading for some of the best views in the park. From the top of the steps, the trail is mostly level, running west along the face of the bluff about 150 feet up from the river bottom. This trail affords grand views of the Lower Wisconsin River valley (Figure 3.9), especially during late fall and early spring when the leaves are down.

At such a viewpoint, stop for a minute and imagine turning the clock back to the time when the glacial wall of ice was sitting to the northeast. You would be standing on a sandstone bluff, just as you are now, but you would see no trees or other vegetation, except for lichens on the rocks around you. The weather would likely be cold and windy. In front of you would be a sea of ice water rolling rapidly to the west. It might even be flowing against the bluff on which you are standing, or at least the broad river would be close. It would extend all the way across to where you can see the row of hills on the north side of the valley.

3.8 Beds of ancient sandstone make up the river bluff at Tower Hill State Park. The shot tower was reconstructed in recent years.

3.9 The Wisconsin River Valley from the bluff at Tower Hill State Park.

The site of Spring Green, the town you now see on the river terrace over there, would be under the roiling icy waters.

Flash forward to the present and you are looking at an expanse of wetland shrubs and trees and other bottomland vegetation. It is growing thickly on silt and sand hauled long ago by the river from the wall of ice to the northeast, which much earlier had hauled this load from farther north and east.

Back on the trail, as the wooden tower above you becomes visible, you'll see a set of steps rising to your left toward the tower. These are broken and rickety and the other end of the trail they lead to is closed off at the top, so avoid these steps. The trail continues curving along the face of the bluff, with more good views of the river valley, and gently descends into the campground near the pavilion where you started. The town of Helena once stood at this site, and a monument now proclaims it to have been "a thriving and important town of lead-mining days." This was where the shot tower workers and their families lived. After nearly 30 years of lead shot production, the process became unprofitable and the shot tower was closed. The new railroad bypassed the town, and shortly after that, Helena was abandoned.

Blue Mound State Park

The thick layer of Silurian dolomite that once covered all or most of Wisconsin was slowly dismantled and carried away by wind, rain, snow, and ice. But remnants can still be found. One such remnant is the top of Blue Mound, the highest point in southern Wisconsin. The dolomite cap on the mound has resisted erosion and kept the mound from being worn down to the level of the surrounding land.

This is the site of Blue Mound State Park. In the early 1900s, the broad, flat area atop the mound was occupied by an oval racetrack.[9] Now, as part of the park, it is a picnic area perched about 415 feet above the base of the mound. Under the feet of picnickers is 100 feet of dolomite lying atop deep formations of sandstone and shale. Deeper still are layers of bedrock rhyolite and granite more than 2 billion years old.

The picnic area has two observation towers. From the east tower, you can look toward the glaciated region. From the west tower, you can view the Driftless Area and see several other mounds, smaller than Blue Mound but having a similar structure.

One of the reasons the mound has survived erosion is the presence of chert, the very hard rock formed from dolomite by the infusion of silica from ancient seawater. Chert was used by Native Americans to make tools. Natives and early white explorers also knew they could make sparks by striking flint rock, and they used it as a fire starter. Chert is the most abundant component of the rock cap of Blue Mound, and over eons, large chert boulders have broken off the cap and now litter the woods around the peak of the mound.

Chert is an interesting rock. The boulders typically have mottled surfaces that splinter when fractured. They are often pocked with holes and small crevices that fill with rain and provide drinking water for birds and squirrels (Figure 3.10). Mice, chipmunks, and squirrels use these pockets in the rock as storage places for their food and even as places to live. Some surfaces on the boulders gleam in the sun, due to the presence of quartz crystals that formed in some of the smaller pockets millions of years ago. Chert can also be marked by bands of color: red, orange, and yellow, caused by the presence of iron oxide, and brown, black, and green, caused by manganese oxide. It is the hardness of this rock that makes the mound.

3.10 Chert is the distinctive rock that makes up much of Blue Mound.

TRAIL GUIDE

Weeping Rock and Flint Rock Nature Trails

Blue Mound State Park has several trails that wind down and around the four sides of the mound. They take you through the shallow valleys where rain, snow, and ice are steadily dismantling the mound, grain by grain. One of these valley trails is the Weeping Rock Trail on the east side of the park. It meanders along a small stream on a one-mile route past a rock wall that is continually washed by groundwater seeping out of its many tiny crevices. The trail is rugged and steep in a few places, but it is worth the hike.

Another interesting hike can be made on trails on the west and north sides of the park. I started near the west observation tower and followed the Flint Rock Nature Trail, which works its way down the west side of the uppermost part of the mound and curves around to the north side. The trail passes a variety of chert boulders and trail signs that explain the geology of the park. As it curves around and ascends the north side of the mound, it crosses the Indian

3.11 Large boulders made of chert and dolomite are found at Blue Mound State Park.

Marker Tree Trail. I took a right onto this trail, which passes through a field of chert boulders of all shapes and sizes. Some of these boulders are huge. Many are partly covered by moss and other plants, and you get a sense of how ancient they are and how they are gradually being broken down by weathering and plant growth (Figure 3.11).

Governor Dodge State Park

At the entrance to Governor Dodge State Park, if you took a quick look around and knew nothing about the landscape, you might think you were in the middle of a broad, rolling, and uninterrupted prairie. As you travel into the park, you quickly realize the land is very much interrupted. Descending from the highland into the valley on the park's main road, you find another world awaiting you (Figure 3.1).

This is the largest state park in the Driftless Area, covering more than 5,300 acres. While other parts of the Driftless are dominated by steep-sided valleys, this park features more of a balance between upland and lowland. Hiking in the park can give you a sense of what the area was like before European immigrants arrived, much of it a broad prairie incised by streams and deep, forested valleys.

The highland bedrock under the park and surrounding farmlands is Platteville dolomite deposited by the Ordovician sea around 460 million years ago. Beneath it are deep deposits of St. Peter sandstone put down 10 million years earlier by streams flowing into the advancing Ordovician sea. As the sea covered the land and became deeper, it deposited the limestone that became today's dolomite bedrock. Since then, seas receded and advanced again several times, putting more layers of sandstone, shale, and limestone over the land, but those succeeding layers have since been eroded away in the park area.

As part of that erosion, ancient streams cut through the bedrock into the St. Peter sandstone, creating the outcroppings you see today in the park. This particular type of sandstone, when it is freshly exposed, is loosely cemented and crumbles easily. With weathering, however, its surfaces become harder. If pieces of the park's bluffs were continually broken off, the bluffs would not stand for long. Those that have had a chance to become weathered and hardened are still standing today.

However, erosion continues. The streams that have most recently been helping to dismantle the sandstone make up the headwaters of Mill Creek, but their erosive work has been slowed by the creation of two lakes. The state built earthen dams across Mill Creek in two places, forming Cox Hollow Lake and Twin Valley Lake, to provide swimming, boating, and fishing for park visitors.

Because sandstone is porous, groundwater seeps through it and has created shallow caves in the park, two of which can be reached on marked trails. Elsewhere, wind, water, and frost have also eroded softer sandstone from under ledges of harder rock, scouring out shallow rock shelters. Pure groundwater flows out of this sandstone body in numerous springs throughout the area. The caves, rock shelters, and springs made the park area attractive to the earliest inhabitants of the region as long ago as 8,000 years. Archaeologists think the park provided wintering grounds for people who fished the nearby Wisconsin and Mississippi Rivers during warmer months. The rock shelters and caves protected them from wicked winter winds and deep snows. The springs, which kept flowing through the winter, provided water for drinking and cooking.

The park area was slightly affected by the lead rush of the 1820s. One of the first deposits explored by European miners was discovered in Cox Hollow a little south of the park and northeast of what is today Dodgeville. Like the park, that town was named for General Henry Dodge, who helped to settle disputes between Ho-Chunk and European immigrant miners over claims to lead deposits. He later became governor.

Most European settlers arriving in the park area came to farm the upland ridges, where the topsoil under prairie grasses was rich with loess blown heavily across the land by postglacial winds. Those farmers also valued the many springs in the park area. They built stone structures called springhouses over the places where springs emerged from the rocks. The cold, fresh water pooled inside the springhouses, which served as cold storage for milk and other perishable foods as well as a source of drinking water. Several of these sturdy structures still exist, and three of them are accessible in the park. (A springhouse tour brochure is available in the park headquarters.)

One of the old farmsteads, the Henry Lawson estate, grew to 160 acres. It was acquired and owned by Iowa County until 1948 when the county gave the land to the state, which made it a state park. Since then, the park has grown to include one of the most diverse collections of landscapes and geological and historical features of all the state parks.

TRAIL GUIDE

Pine Cliff Trail

This 2.5-mile hike takes you to a pine-studded promontory overlooking the confluence of two stream valleys, now partly filled by Cox Hollow Lake. It is moderately difficult with a few steep sections, some slippery spots in the wet areas, and a total elevation gain of 223 feet. The trail begins at the Enee Point picnic grounds, crossing a stream and following it for a tenth of a mile. The high ridge on your right is made of bedded St. Peter sandstone. You cross the stream again on a bridge to begin the hike to the top. It is fairly steep for another tenth of a mile and then evens out to a gentler rise. To your left, look for breaks in the foliage to get a view of the top of Enee Point and the ridge running back from it. About 400 million years ago, you would be standing on a flat expanse of rock covering the whole area, but over many millions of years, the land between you and Enee Point has eroded away.

3.12 One of the cliffs for which the Pine Cliff Trail is named, overlooking Cox Hollow Lake in Governor Dodge State Park.

A third of a mile into the hike, about halfway to the top of the ridge, is a bench and a sign explaining the series of ancient horizontal sandstone arches exposed in several places. They show how some layers of sandstone are harder than others. Layers bonded by iron or calcite erode more slowly, so the softer layers above and below them erode into the bluff. Continuing up the moderately steep steps, you pass several of these curved shelves of rock. The little stream that crosses them has carved them over time, creating the shallow gorge through which you are climbing. A boardwalk crosses the stream over one of the larger arches.

At the half-mile point is a trail junction. The trail to the right runs across the ridge going southeast. The left-hand trail continues on a mostly level east-northeast route toward a pine-covered cliff. At the one-mile point, if you are up for some climbing, you can scramble up a 15-foot rock outcropping to a short, level spur trail to the promontory overlooking the north branch of Cox Hollow Lake (Figure 3.12)—one of the park's most beautiful vantage points.

Back on the main trail, you hook to the right and proceed along the south side of the ridge, overlooking the south branch of Cox Hollow Lake. After about a tenth of a mile, the trail goes toward the lakeshore directly across from cliffs on the other side of the lake. You get good views of this spectacular vertical sandstone cliff at several points on the trail as it hugs the lakeshore for about a quarter mile.

From the point where the trail rises away from the lake, it is a quarter mile to the next trail junction. After leaving the lakeshore, the trail drops fairly steeply down through the area where the tornado hit in summer of 2014, causing massive tree damage. The Pine Cliff Trail continues south from this area and loops around to the other side of the lake. I took a cutoff trail that goes northwest across the ridge and back to the Enee Point parking area. It begins with a fairly steep tenth-mile hike, then levels off to a gentle downslope leading to the trail junction that closes the loop around the ridge. From here, you can return to the parking area on the trail you took up the ridge to this point.

TRAIL GUIDE

Stephens Falls and Lost Canyon Trails

On this hike, you can travel back in time by walking through a canyon carved over millions of years by water, wind, and ice. As you hike into it, you are passing through thousands of years of sandstone deposition for every foot you descend. Ironically, this is one of the park's canyons that was not lost; others disappeared when they were flooded to create lakes.

You access the Stephens Falls Trail from the parking area on the west side of the park. The trail passes by the well-preserved Stephens springhouse that served as a natural refrigerator for the Stephens farm family for decades. It then takes you into the canyon on stone steps that are rugged but equipped with a sturdy railing. (If you do not care to take the steps down, you can bypass them on the Lost Canyon Trail, which splits off near the trailhead, skirts the area on high land, and descends more gently into the canyon. However, I strongly recommend the Stephens Falls Trail for the best views of the canyon.)

As you descend, the ancient beds of sandstone are clearly visible in the canyon wall on your left. This route offers great views of Stephens Falls, a 30-foot waterfall on an unnamed stream that has carved a wide arch into a sandstone wall (Figure 3.13). The ledge of harder sandstone has been eroded back by the stream for thousands of years, while the softer sandstone has been scoured away from under the ledge. From the foot of the steps, you can walk to the base of the falls. The trail follows the creek downstream, passing between high sandstone cliffs and rock outcroppings. Old white pines cling to cliffs on either side, and many boulders on the canyon floor are dressed in thick moss.

The Stephens Falls Trail runs through a tributary canyon that merges with the wider valley called Lost Canyon. About a quarter mile from the falls, the trail ends at the junction with Lost Canyon Trail. I took a left at the junction and continued south-southeast. From here, the trail is gentle, wide, and smooth. It stays close to the right side of the valley, passing more striking cliffs and outcroppings. To the left, the stream has soaked the valley floor, which has gathered vegetation and become a small wetland. The drier areas of the valley floor are studded with big old white pines.

About a half mile from the trail junction, the trail crosses the wetland on a dry path over a stream culvert and ascends the ridge on the east side of Lost Canyon. On the trail to the top of the ridge, the dark-colored soil underfoot is something between sand and clay. It probably contains loess, the silty material that was blown across the land during the retreat of the glacier. It is what made the uplands of the park fertile for the lush prairie and later for farm fields.

The trail rises along the ridge and about a quarter mile from the culvert makes a sharp turn to the left. From the beginning of this turn to the ridgetop

3.13 Stephens Falls shows how water erodes softer sandstone layers under harder ones, creating a horizontal arch in the canyon wall.

is the only steep section of this trail. As it levels off, it heads northwest along the ridgetop heading back toward the trailhead. Along this section, you can see the origins of the streams that are carving the valleys, as spring-fed trickles emerge from the forest floor and work their way down into the canyon. One such place is Wilson springhouse off the trail a little to the left about a half mile from where the trail topped the ridge. On the early spring day when I hiked the trail, the stream formed by the springs there was choked with watercress—the season's first bold splash of green decorating the woods.

As the trail proceeds, it runs between the canyon and a field to the right—presumably once one of Farmer Wilson's fields, now a prairie restoration project. Eventually, this area will look much as it did thousands of years ago after the glacier departed and prairie grasses gradually took over the open uplands.

As the trail nears the trailhead, it skirts the rim of the canyon where this hike began. About a quarter mile from Wilson springhouse is a gorgeous overlook of the canyon floor just downstream from Stephens Falls.

Not far from this overlook, the trail curves sharply left around the top end of a steep wooded valley, a common feature of the Driftless Area still being carved by rain and small streams. Thousands of years from now, it will be a steep tributary valley merging with Lost Canyon.

TRAIL GUIDE
Deer Cove Rock Shelter

On the road to Cox Hollow, between the boat landing and the picnic area, is Deer Cove, where more than one group of Native Americans found shelter under a rock overhang beginning perhaps 7,000 years ago. It is just a quarter of a mile from the parking area to the overhang—a 35-foot-high ledge of resistant sandstone over a more eroded wall of weaker sandstone.

A trail sign near the parking area informs visitors about what archaeologists have learned from Native American artifacts found at the shelter. The overhang also tells a fascinating geological story. The tall vertical sandstone formation beneath the ledge (Figures 3.14 and 3.15) was formed from sand dunes that accumulated millions of years ago near the edge of a sea. In the park area, this sandstone averages 100 feet thick. The flatter, harder layers of stone in the overhang were laid down by an Ordovician sea that advanced over the area, drowning the high sand dunes.

3.14 This overhang sheltered Native American groups during many winters in the distant past.

3.15 Close up, you can see the unique quality of the sandstone that was formed from sand dunes over millions of years.

To the left of the area shown in Figure 3.14, the ground rises to a level much closer to the overhang. The climb is rocky and steep in places, so be careful, but you can get close enough to examine the overhanging layers of sandstone. You will see small vertical crevices and thin ribs of rock running up through these sandstone layers. These were likely made by worms that lived on the shallow sea bottom around 450 million years ago.[10]

Wyalusing State Park

This park is famous for its grand views of the confluence of the Wisconsin and Mississippi Rivers. It sits atop bluffs composed of Ordovician dolomite over Cambrian sandstone. The dolomite layer is thicker here than in other parks, which is what makes this area's geologic story unique.

Recall from Chapter 1 that about a billion years ago, parts of North America were warping up and others were warping down, and one result was the Wisconsin Arch. At the end of this process, layers of sandstone and dolomite that had been deposited on flat sea bottoms were mounded up in northern and central Wisconsin so that they were slightly sloping on the east and west sides of the state. Now imagine a giant sanding machine pressing down on the state. The highest points would be sanded off first, and around the perimeter of the mounded area, we would see the layers of underlying rock exposed in cross-section.

Erosion worked in much the same way, operating over 400 million years to scour off most of the upper layers of dolomite. Because the dolomite layer atop the Wyalusing State Park area was sloping southwest, more of it remains in the face of longtime erosion. Thus the dolomite in its bluffs is thicker than it is in the bluffs to the north and east, upstream on the Mississippi and Wisconsin Rivers.

Being in the Driftless Area, Wyalusing State Park was never directly marred by glaciers. However, geologists found evidence of some ice coverage near the mouth of the Wisconsin River just below the park's bluffs. Deposits of till many hundreds of thousands of years old were found on the north side of the river east of where it joins the Mississippi. The researchers thought a lobe of ice from an earlier Pleistocene glacier must have invaded the area from the west, plugging the mouth of the Wisconsin River. As this lobe melted, water flowed eastward in the river valley (opposite today's direction of flow), leaving ancient upstream-oriented outwash in parts of the lower Wisconsin valley.

However, the more recent research by Carson and colleagues (see pages 68–69) offers a different explanation for this outwash. The Wisconsin River valley is quite different from most river valleys. Rivers generally start out small and grow as tributaries feed them, and their valleys generally widen steadily from source to mouth. The Wisconsin River valley, however, narrows dramatically as it approaches its mouth (Figure 3.2). Upstream about 90 miles at Sauk City, the valley is 4 miles wide; at Bridgeport, near the river's mouth, the valley is just a half mile wide.

The traditional explanation for the river's downstream narrowing is that the river had less trouble cutting into the sandstone bedrock upstream than it had wearing its way through the thick dolomite layer at the river's mouth. Carson's hypothesis explains this narrowing more effectively, however. If the river once flowed east in the lower Wisconsin Valley, it would have widened from west to east, just as it does today, and the ancient upstream-oriented outwash would have been deposited by that east-flowing river.

Regardless of how the river flowed before the Wisconsin glaciation, it was certainly flowing into the Mississippi afterward. The point where the two rivers meet must have been quite a sight, with tremendous volumes of water merging there, creating the massive flow that would course south across the continent.

With the end of the postglacial drama, the climate warmed and life returned to much the same condition it was in before the glacier came. Waters flowed across the park area and sank into the ground and one effect was the establishment of karst topography, a type of terrain featuring caves, sinkholes, and springs. Karst exists wherever bedrock is soluble, as the dolomite underlying Wyalusing State Park is. Over time, water can dissolve the calcium in this rock, creating crevices and caverns, and thus the park encompasses a number of caves, including Treasure Cave (Figure 3.16). It can be reached on a short but rugged spur trail that branches from the Sentinel Ridge Trail, then ducks through a narrow keyhole opening before ascending a ladder to the cave.

Other caves were eroded from the St. Peter sandstone that underlies the dolomite in the park. It is a loosely cemented stone that was deposited on an Ordovician seashore and blown by winds into deep sand dunes. It is almost pure quartz but sometimes contains iron, which gives it a reddish color, as in the park's Sand Caves. Because St. Peter sandstone is easily eroded, these caves were created by wind and water action, not dissolved from harder rock as Treasure Cave was.

The first humans in the Wyalusing area arrived about 11,000 years ago, hunting big game in the Paleo tradition and following the retreat of the glacier. Much later, people of the Hopewell and later cultures built burial mounds and effigy mounds for religious or ceremonial purposes. Many of these can be viewed in the park, especially from Sentinel Ridge on the Mississippi bluffs.

Several areas of the park are named for the members of the Ho-Chunk and other tribes who lived here over the centuries. Eagle Eye Bluff and Yellow Thunder Point are named for prominent leaders, and Signal Point was the site of signal fires. In the nineteenth century, a tribe of Munsee-Delaware Indians settled in the area. The word *Wyalusing* has been traced to this Lanape-speaking group and translates to "home of the warrior." The last Native Americans to live in the park were Ho-Chunk camped on Sentinel Ridge under the leadership of Chief Green Cloud. In 1882, to satisfy the terms of an 1829 treaty, Chief Green Cloud led his people out of the area and across the river into Iowa. The Native

3.16 The view from inside Treasure Cave looking out at the Wisconsin River Valley, Wyalusing State Park.

American history of the park area is documented on a remarkable trail sign near the Green Cloud Shelter and picnic area overlooking the Mississippi River valley. It provides detailed information, including exact locations of camps, mounds, and trails from the Mississippi to the Madison Four Lakes area.

Because at least 14 different tribes were known to have lived in the area or regularly passed through on trade routes, archaeologists presume they considered the confluence of the two big rivers to be neutral territory. Not so for the Europeans who arrived after the first French explorers and fur traders came, following the lead of Father Jacques Marquette and Louis Joliet, who first paddled through the confluence in 1763. Eventually, the French and English fought to control the confluence. The bluffs of Wyalusing made a strategic site for those vying for control of the lucrative fur trade.

Miners were also interested in the area, knowing of rich lead deposits to the south. Some efforts were made to mine lead within the present-day park boundaries, but records do not show any successful mining operation. Farmers, however, had better luck growing crops on flat ridgetops and valley floors where the postglacial accumulation of loess had enriched the soils.

In the late 1800s, the Robert Glenn family owned the land under Wyalusing and wanted it to become parkland. The Wisconsin legislature approved the purchase of the land in 1912, and the park was established in 1917. In the 1930s, the CCC was employed to build some of the park's roads, trails, fireplaces, shelters, and picnic areas.

In addition to being a prized destination for hikers, canoeists, and campers, Wyalusing State Park is also valued highly by ecologists, archaeologists, and geologists. The park's Mounds Archeological District is nationally recognized and used as a model for research. The northeasternmost section of the park is designated as the Wyalusing Hardwood Forest State Natural Area. It preserves four major forest types and two major soil types developed in loess, all of interest to ecologists. It also includes dry bluff-top as well as floodplain ecosystems that serve as ecological reference areas. Rare birds living in these woods include Kentucky warblers and Acadian flycatchers. The area contains no marked trails, picnic areas, or other developments. But even the developed areas of the park are great places to research, teach, and learn. Many school teachers have taken their students on highly rewarding field trips to the park.

TRAIL GUIDE
Old Wagon Road Trail

This trail is rugged because of some steep sections, although most of it is fairly level. It passes through classic Driftless topography: a deep coulee carved into ancient layers of rock over hundreds of millions of years by wind and water. Now vegetation is also helping to dismantle the canyon walls.

Find the trailhead at the Knob Overlook and Shelter at the north end of the park. Before going to the trailhead, look east from the Knob parking area and take some time to view the valley of the Wisconsin River (Figure 3.17). From this point 10,000 years ago, you would see a roiling bluish-gray flood of water spanning the valley and flowing much higher than today's water level. There would be no woods around you, only lichens and mosses clinging to rock outcroppings. The postglacial landscape would be wet, rocky, and very cold.

3.17 The Wisconsin River Valley as viewed from the Knob Overlook at Wyalusing State Park.

The trail first descends steeply past the Knob Shelter angling down the bluff side toward the south bank of the Wisconsin River. A quarter mile from the parking area, it turns right to parallel the river, running along the base of the ridge on which the shelter is built. It rounds the eastern end of the ridge and follows it back into the coulee formed by a small seasonal stream. The trail descends further to the streambed and crosses it at the 0.6-mile mark. From there it joins the old wagon road for which it is named, angling up the other side of the ravine just crossed.

Hoofing up this rather steep old road, you pass remarkable outcroppings of Cambrian sandstone created on an ancient sea floor 500 million or more years ago. Some of these sandstone beds are visibly tilted, reflecting some sort of upheaval that occurred long after they were deposited. The road runs from the top of the bluff down to the river to a landing called Walnut Eddy. It is such a steep hike that I found myself pitying the poor oxen that had to pull loaded wagons up the steady incline. Similar sympathy could be felt for the poor drivers who had to control the wagons as they rolled down this road to pick up another load. This section of the trail rises to the point where the canyon narrows, shallows, and disappears on the forest floor where the streambed originates. At the 1.5-mile mark, the trail ends at the picnic and recreation area on top of the ridge.

4.1 Devil's Doorway, an iconic rock formation on the East Bluff at Devil's Lake State Park

4

Carved by Water and Ice

South-Central Wisconsin

About 1,700 million years ago, southern Wisconsin lay within a tropical rocky, sandy plain. Intense winds scoured the barren Penokean Mountains to the north, and tropical rains washed sand and clay from these highlands into streams that flowed south to a sea that lay over what is now Illinois and Iowa. These were braided streams—broad and multichanneled—and they spread the sand and other sediments far and wide.

Upon the rhyolite-granitic bedrock, these streams and the advancing sea deposited first a layer of clay-rich soil and then tremendous amounts of sand and fine gravel. As the sand layers deepened, the sand grains were cemented into sandstone by solutions of quartz and other minerals. Over tens of millions of years, the sandstone layers reached depths of up to 4,000 feet. Under the sandstone, the clayey layer had been converted to shale.

This long period of quiet building came to an end sometime between 1,650 million and 1,450 million years ago, when the primitive North American continent, driven by tectonic forces, collided with a smaller continent to the south. Over tens of millions of years, the land in the coastal collision zone was crumpled and folded in complex ways, and the rock layers beneath were churned and metamorphosed by heat and pressure. The sandstone was forged into the hard rock quartzite, while layers of shale became slate. Underlying masses of rhyolite bedrock were heaved to the surface in some areas. Under the surface, magma was moving into chambers opened by the collision, creating new masses of granite within the mix of bedrock. The end result of this process was a low mountain range underlain in places by granite and composed of quartzite and rhyolite. It stretched from southern Wisconsin to the west, possibly as far as the Dakotas.

Of particular interest to geologists around the world is one area of crumpled land called the Baraboo Hills, located near the center of the southern third of Wisconsin (see the area around Devil's Lake Park in Figure 4.2). It is shaped something like a rowboat pointed east, the boat's gunwales forming three ranges of hills, and the boat's interior forming a central lowland.

The crumpling force came from the north, like a huge foot pushing one end of a rug toward the other end, creating folds in the rug.[1] On the north side of the boat-shaped formation, this folding action raised the rock layers to a nearly vertical position. South of that range of hills, the land was depressed into a trough formation (the interior of the boat), and the range of hills to the south was raised much less steeply—folded less sharply—than the north range. To the east, the two ranges came together at a point, like the bow of the boat. To the west, a north-south range of hills formed the stern of the boat.

Geologists refer to the central part of this formation, where roughly parallel ranges of hills were raised by the crumpling of once-flat land and the area between was depressed, as a syncline. The fact that the west range of hills runs north and south, perpendicular to the syncline, shows how complex the crumpling of land can be when continents collide.

As the mountain building came to an end and the Precambrian sea receded, the steady forces of erosion went to work and, over the next 500 million years, leveled the land once again to a rolling rocky, sandy plain. In south-central Wisconsin, the more stubborn quartzite peaks of the Baraboo Hills resisted erosion. At the beginning of Cambrian time, they stood about twice as high above the plain as they do today. Geologists Robert H. Dott Jr. and John W. Attig provide the best description of south-central Wisconsin at this time:

> The region must have been a pretty strange place 500 million years ago . . . a sandy, windblown desert almost barren of life, but not dry like deserts today. . . . There were temporary ponds, streams, and wet sand flats among the dunes. There were no trees, shrubs, or grasses. Only lowly microbes had learned how to live above sea level, although . . . a few small crablike animals were beginning to experiment with occasional walks on the moist beaches and dune flanks.[2]

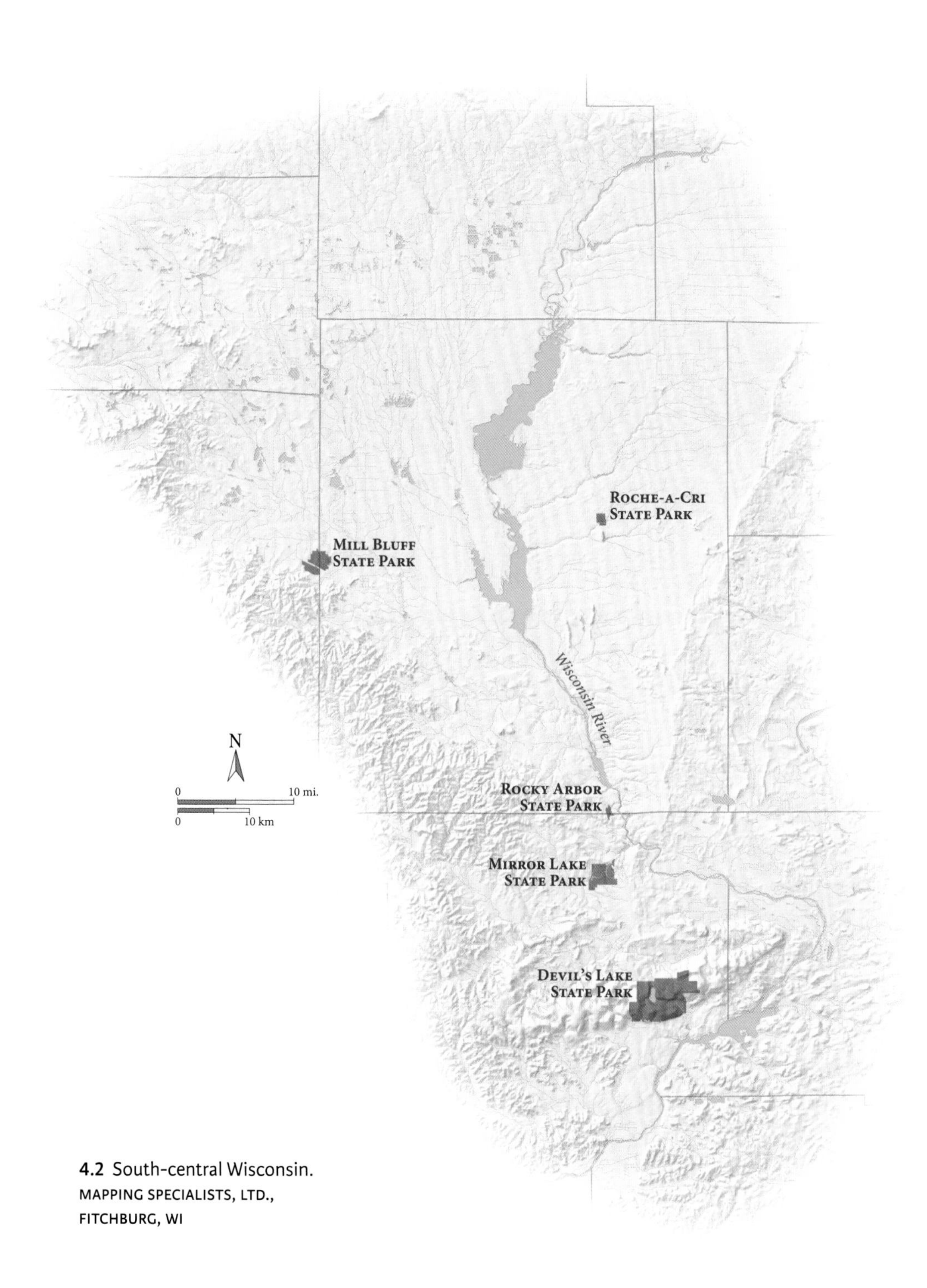

4.2 South-central Wisconsin.
MAPPING SPECIALISTS, LTD., FITCHBURG, WI

When the Cambrian sea first arrived, gusty winds and rivers flowing to the sea brought sand and clay. These braided streams and the encroaching sea itself slowly filled lowland areas with sediments that would become sandstone. When the shallow Cambrian sea first covered south-central Wisconsin, the Baraboo Hills and other quartzite peaks in the area were rocky islands often battered by waves driven by tropical storms. Geologists think that the Cambrian seas stood long enough in the area to completely fill the lowlands, including the one lying between the two ranges of the Baraboo Hills.

The withdrawal of the Cambrian sea brought a long period of erosion followed by the advances and retreats of subsequent seas, time and again through Ordovician, Silurian, and Devonian time. With each advance came deposits of sandstone, shale, and limestone, or dolomite, followed by a period of erosion. Residual masses of sandstone and limestone found on the tops of some of the higher peaks show that the entire area was once buried by the younger rock that formed the ancient sea beds. After the inland sea retreated for the last time, wind and water had hundreds of millions of years to wear away most of those softer layers of rock. Against such unrelenting forces, though, the hard quartzite peaks themselves persisted and still do today. Geologists speculate that one day in the distant future, continuing erosion of softer rock will again leave those resistant peaks standing twice as high as they are now above the surrounding lowlands.

The Ice Age brought a new form of erosion, more intense and less forgiving than wind or rain could ever be. Masses of ice simply plowed up vast areas that had been delicately carved for eons by gentler forces of erosion. South-central Wisconsin is interesting because it exhibits landforms molded by the glaciers lying side by side with land never touched by them. Still, many of those untouched areas were profoundly affected by the ice masses.

We know little about the effects of glaciers prior to the most recent one, which covered most of Wisconsin between 30,000 and 11,000 years ago. The segment of that glacier that covered part of south-central Wisconsin was the Green Bay lobe—a vast tongue of ice that flowed down the lowland that now contains Green Bay and Lake Winnebago. (You can see its outline in Figure 1.1, bounded on the east by the Door Peninsula and on the west by the brownish highlands angling from northeast to southwest.)

The lobe's major thrust was to the south, but at its western margin, it was pushing ice directly west. It crawled up the eastern flank of the Baraboo Hills and covered some but not all of those peaks. It stopped advancing at a line that crosses the Baraboo range about halfway between its east and west ends (Figure 4.2). There it sat for many centuries, building a terminal moraine along that line.

When the glacier arrived some 19,000 years ago, the Wisconsin River was flowing around the east end of the Baraboo Hills, just as it does now. The ice wall inched toward the hills from the east and eventually blocked its flow as it made contact with the easternmost hills. This event led to a dramatic change in the character of a large area of south-central Wisconsin. The damming of the river created Glacial Lake Wisconsin (Figure 4.3), a vast shallow expanse of ice water that would grow to 1,825 square miles—larger than today's Green Bay and nearly as large as the Great Salt Lake in Utah.

As the river's water spread north and west into the Driftless Area from the site of the glacial dam, it flowed into valleys and rose among ridges that must have looked like softer forms of the ridges and hollows of today's Driftless Area. It lapped against the row of higher hills to the southwest and reached a maximum depth of about 150 feet near its southwest end. The glacial lake rose to the point where its waters could exit the basin on its north side through what is now the east fork of the Black River. Within the flooded area, higher bluffs made of harder forms of sandstone, some of them capped by dolomite, had resisted the erosion of the centuries and now stood as islands in the lake.

The vast lake sat in the glacial deep freeze for at least 3,000 years while streams brought sand and gravel from the ice masses to its shores and winds blew dust across its surface. These sediments, along with tons of sand eroded from the sandstone islands by lake waves and currents, settled to the bottom and gradually filled the drowned valleys, creating a level lake bottom. By about 14,000 years ago, the climate was warming and the glacier was melting back. Meltwater streams flowing from the ice walls increased in volume and dumped more sand into the big lake. Geologists estimate that the lake bottom sediments are as much as 300 feet deep.

When the Green Bay lobe began its retreat, it left a ridge of sand and rock—the Johnstown Moraine—along most of its maximum extent. A part of this ridge served as the eastern shore of Glacial Lake Wisconsin (Figure 4.3). Southeast of the lake, to the north and south of the Baraboo Hills, new glacial lakes formed between the Johnstown Moraine and the retreating ice wall. The lake north of the hills filled the Lewiston Basin to the level of Glacial Lake Wisconsin. The southern lake, called Lake Merrimac, was the ancestor of today's Lake

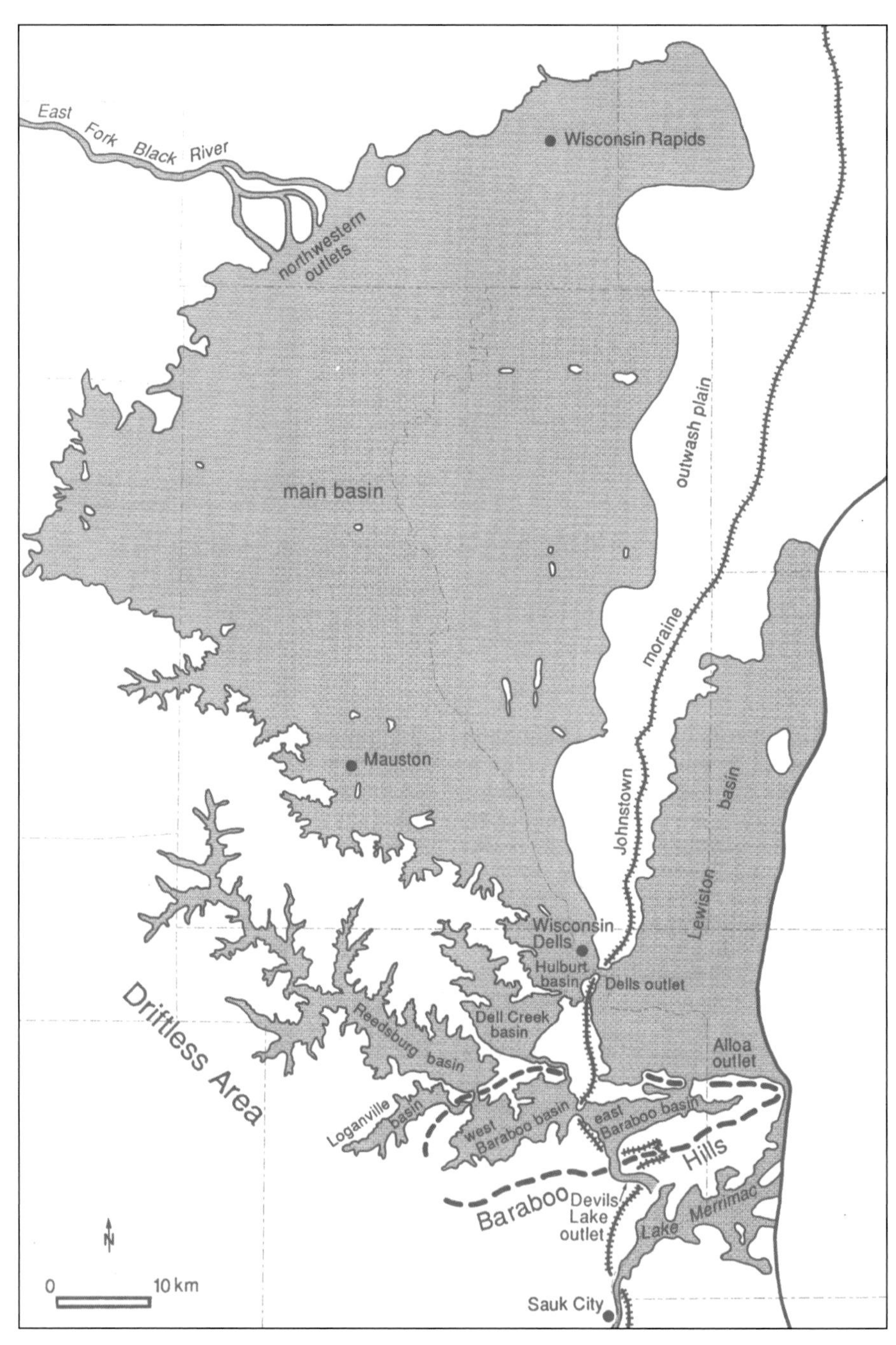

4.3 Glacial Lake Wisconsin included a main basin and several smaller extensions. CLAYTON AND ATTIG, *GLACIAL LAKE WISCONSIN*, GEOLOGICAL SOCIETY OF AMERICA

Wisconsin. Other, smaller lakes formed in the center trough of the Baraboo Hills (Glacial Lake Baraboo) and in deeper valleys to the northwest.

The slow formation of these lakes was the quiet precursor to the next geological event that would change the south-central Wisconsin landscape. The lake in the Lewiston Basin was held back at its southern end by glacial ice, and with the climate warming, this ice dam's days were numbered. When the dam broke, the water from Lewiston Basin gushed into Lake Merrimac, which rose quickly and overwhelmed the Johnstown Moraine at its western end. It cut a notch through the moraine and began the rapid carving of the Wisconsin River valley downstream from that point.

To the north of the Baraboo Hills, all of Glacial Lake Wisconsin was held back by the sand and gravel ridge of the Johnstown Moraine, which had been built on loosely cemented Cambrian sandstone. As the glacier retreated and the land thawed, these barriers became weaker. When the level of the Lewiston Basin lake dropped an estimated 100 feet in just a few days, the moraine dam could no longer hold back the tremendous volume of Glacial Lake Wisconsin's water. It broke through the dam and the resulting torrent was a force of nature. At its main drainage point, near today's Wisconsin Dells, it quickly stripped away lake bottom and glacial sediments and began carving up the underlying sandstone. This flash flood was so powerful it rolled boulders as big as five feet in diameter for miles down the Wisconsin River. Geologists estimate that within a few days or weeks, the main body of the lake and all the nearby lesser glacial lakes had drained away.[3]

Within a heartbeat of geologic time, the vast body of water that had lain for at least 3,000 years in the central part of Wisconsin was gone. Within a few weeks, the famous gorge of the Wisconsin Dells was created, along with similar dramatic features that we see in some of the area's state parks. The torrent of icy lake water had hauled gigantic amounts of sand and gravel down into the Wisconsin River valley. And to the northwest lay a vast mucky plain where the lake had been, its rocky islands now standing mute as scattered spires, crags, and buttes.

Geologists think that lesser versions of this flooding and draining of glacial lakes could have occurred more than once. However, when the glacier finally retreated, the next phase of the south-central Wisconsin story was a long period of warming and drying of the land. High winds blew the sand and silt across the old lake bottom and formed dunes in some places. Hardy pioneer plants gradually took hold on the sandy expanse, and the shrub lands and forests that cover much of the area today became established.

To the south, in the Baraboo Hills where the topography was anything but flat and unvaried, a larger variety of ecosystems took hold after the glacier was gone. Upland prairies developed next to oak savanna that bordered on thick forests descending into the valleys. In the deeper hollows, the cooler climate favored species normally found farther north, such as hemlock and yellow birch. This sort of diversity, including a large variety of woodland and prairie bird species, can still be found today in the less-developed areas of the state, such as the state parks.

The Baraboo Hills always have drawn travelers, starting around 12,000 years ago when the first Paleo-Indian hunters came to the area. Between 1,400 and 800 years ago, mound-building tribes lived in south-central Wisconsin. Effigy mounds have been preserved in some of the area's parks, as have petroglyphs, or carvings on rock outcroppings made by Native Americans in the same time period. These are records of hunting and other activities that were part of life in the Baraboo Hills and north of the hills on the central sand plain.

As in any area, the geology of south-central Wisconsin has helped to determine the uses of the land. In the Baraboo Hills, early hunters roamed among spruce, aspen, and ash in the highlands looking for mammoths, mastodons, and other big game. These woodlands were the first to become established after the glacier. Within 2,000 years of the glacier's retreat, birch, red pine, and white pine became more dominant, supporting another community of animals as human hunters shifted to smaller game, especially deer. As the climate warmed, plant communities became more diverse, including maple, basswood, and elm trees. Within the past 5,000 years, oak forests have become prominent.

In the Central Sand Plain north of the hills, a great deal of sand and silt washed away from the retreating glacier in meltwater streams. These streams left broad, fan-shaped deposits of outwash sediments from the east side of the glacial lake bed toward its center. Wetlands formed in the flatter areas as the glacier retreated. Over thousands of years, peat beds formed in these wetlands, and people have used this peat as fuel for heating and cooking and more commonly as a soil conditioner. Sphagnum moss, another popular gardener's resource, also grows in these wet areas and has been harvested widely from this part of the state. And the marshy, flat land has proven ideal for raising cranberries, which have become a major industry in central Wisconsin.

In the drier areas of the central plain, sandy, loamy soils developed. After early immigrant farmers cleared the thin vegetation atop these soils to plant crops, windstorms would often blow fragile topsoil away, creating dunes that can still be found today.[4] Farmers learned to plant rows of evergreens as windbreaks to control such erosion. To grow common grain and vegetable crops, these quickly draining soils need a lot of irrigation and fertilizers. With effort and investment, farmers have made the ancient glacial lake bed a productive agricultural region of the state. However, the unique natural character of the sand plain, and that of the hills to the south, has been well preserved within a number of popular state parks.

Devil's Lake State Park

From all along the East Bluff Trail in Devil's Lake State Park, you can get spectacular views of the deep valley that contains Devil's Lake. One such view is at a point overlooking the south end of the lake where you can take in the entire gorge. To your right, the segment containing the lake stretches away to the north. Below you lies the sweeping broad bend to the left away from the lake and into the wide forested part of the valley that lies to the east. It is one of the grandest of the many views in this gorge that draws 1.5 million visitors every year.

If you could turn back the clock about 2 million years and stand at that point, the rocky cliffs and forests around you would look much the same, but the scene below would be strikingly different. Instead of gazing at a placid lake and a broad beach 500 feet below, you would be looking 800 feet down at a river flowing in the bottom of a V-shaped gorge. The valley would be wide open, not blocked at either end by land masses as it is now. The stream would be sweeping down from the north, swinging lazily around the bend below you, and flowing east out of the gorge to where it would again veer south.

But turn the clock back much further, for the story of this gorge begins with the collision between two continents 1,650 million years ago. The subsequent churning of rock forged the Baraboo Hills, the southern range of which would later be notched by an ancient river. When the Cambrian sea first arrived some 500 million years ago, the quartzite hills were standing probably 1,000 feet above the surrounding barren, rocky plain of southern Wisconsin.[5] Quartzite is so hard that those hills probably have not since been worn down much by erosion. Their reddish and purplish coloring probably also has not changed. It is due to thin films of iron oxides that got mixed with the sandstone as it was converted to quartzite.

For the next 100 million or more years, Cambrian and later seas advanced again and again, eventually filling the valleys among the hills with sedimentary rock and finally burying the entire range. Long before they were buried, however, the bluffs of the Devil's Lake gorge stood as islands, their cliffs often slammed by tropical storm waves. Evidence of such sea cliffs and what the waves did to them can be found on some of the park's hiking trails.

When the seas finally retreated for the last time between 260 million and 300 million years ago, erosive forces went to work on the sedimentary rock mantle that covered much of Wisconsin. In the Devil's Lake area, a network of streams flowed and carved shallow lowlands deeper, and over millions of years, the deepening river valleys channeled faster flows that in turn eroded the limestones and sandstones more quickly. Eventually, the quartzite bluffs emerged once again, standing up to the wind and water that had carried the overlying sedimentary rock away.

Evidence shows that some streams must have been fast-flowing and powerful enough to drill potholes in the quartzite. An eddy in such a stream can capture gravel and rocks that become grindstones whirling against the bottom of the stream. With enough power and time, an eddy can carve a pot-shaped hole into even the hardest stone. (See the description of Interstate State Park in Chapter 2 for a more detailed explanation of potholes.) At Devil's Lake, some potholes were formed high up on the bluffs, which indicates that they must have been drilled when the streambed was at that level, long before the sandstone was washed out of the gorge. By the time the glacier arrived, ancient rivers had carved the 800-foot-deep gorge that would later hold Devil's Lake.

The Green Bay lobe of the most recent glacier pushed ice into the gorge from the east about 19,000 years ago. During the next 4,000 years, the ice crept over the land and up the easternmost slopes of the Baraboo Hills. The ice could not overtop the highest peaks just east of the Devil's Lake gorge, so that area remained free of glacial drift. North and south of that highland, the ice kept pushing westward into the lowlands until it reached the gorge and there it stopped, plugging both ends of the gorge. Apparently, that was when the warming started. Had it begun a few decades later, the ice might have filled the gorge and covered the high peaks around it. Because it did not, the western third of Devil's Lake State Park is in the Driftless Area.

The ice wall of the Green Bay lobe curved through the eastern part of what is now the park and stretched away to the north and south. It stood there for hundreds to thousands of years, while the ice mass pushed huge amounts of sand, gravel, and boulders to its eastern margin, building the Johnstown Moraine, the long, nearly continuous ridge that averages 60 feet in height. It outlines the Green Bay lobe and is easy to see from a satellite view (Figure 1.1).

Glacial ice did not pulverize quartzite as it did some other forms of stone, but the glaciers had an indirect effect on the quartzite hills—more like that of a pickax than a bulldozer. When the frigid weather of the Quaternary period arrived, ice formed in cracks in the quartzite cliffs. With each freeze-thaw cycle, the ice widened these cracks, and over time large chunks of quartzite were pried away from the cliffs and pulled down by their own weight. The result is the immense piles of quartzite blocks, some larger than 10 feet on a side, draped along the bases of the cliffs around Devil's Lake. These piles are called talus (Figure 4.4).

4.4 This talus slope lies below the south bluff at Devil's Lake State Park.

Geologists think the talus was formed during the time of the glacier because no talus sits on top of the moraines left by the glacier. Instead, some talus is thought to be buried by moraine, which means some of it was formed before the ice reached its farthest extent. Geologists also point to glaciated quartzite hills close to the gorge that have no significant talus (such as the slope on the right side of this book's front cover image). Presumably the talus that formed there as frigid weather closed in before the ice arrived would have been moved along by the glacier and now are part of the terminal moraine.

The freeze-thaw pickax effect is also responsible for some of the most popular rock features in the park, including the Balanced Rock (Figure 4.5) and the Devil's Doorway (Figure 4.1 and the Trail Guide later in this chapter). Blocks of quartzite from all around these formations must have been fractured, pried away, and knocked down by other blocks falling from above. These seemingly precarious formations were left standing as giant rock sculptures.

Devil's Lake formed sometime after the glacier advanced—a glacial lake sitting between two walls of ice possibly 200 feet high. It rose to 90 to 150 feet above its present level and found an outlet at the northwest corner of the basin. Its water flowed into Glacial Lake Baraboo, an arm of Glacial Lake Wisconsin to the north. Over time, icebergs calved from the glacier's walls and plunged into the lake. We know this because glacial erratic boulders have been found in the small valley at the southwest corner of the lake where Messenger Creek now flows in. Icebergs must have carried them there, because no glacier covered that area. Divers now also find erratics on the lake bottom where they fell when their ice rafts melted away.

The glacier left two high moraines spanning the gorge at either end of the lake. They are actually connected as parts of the Johnstown Moraine (Figure 4.3). The park's nature center is built near the crest of the north moraine. The south moraine rises away from the south beach area, and its eastern flank overlooks Roznos Meadow, a broad outwash plain. The Roznos Meadow parking area on County Highway DL provides a view of this moraine, which rises abruptly to 150 feet above the meadow like a great earthen dam in the distance.

As the glacier retreated, its meltwater washed huge loads of sand and gravel into the lake's basin from both ends, half filling it. Geologists estimate the depth of this outwash on the lake bottom to be 300 to 350 feet.[6] The lake level is stable even though it has no outlet streams. So because the lake is perched on porous outwash and sits higher than other nearby lakes and streams, it is likely that

4.5 The Balanced Rock formation overlooks the south shore of Devil's Lake.

water seeps out the lake bottom and finds its way through groundwater to these other waterways.

As the ice began to melt and retreat, small glacial lakes formed in several locations in the unglaciated highlands east of the lake, including Steinke Basin, Feltz Basin, and Roznos Meadow, all accessible by car and/or trail. These areas are covered with a layer of lake bottom sediments under the grassland and savanna communities that occupy them now.

Most of these flatlands were farmed in the nineteenth century by families whose names are now preserved in park lore, including Steinke, Feltz, and Hanson. They raised sheep and cattle along with garden vegetables. In winter, some of these farmers went down to Devil's Lake to cut slabs of lake ice, which they stored in ice sheds, insulated with layers of sawdust. It was used throughout the year to keep perishable foods cold in farmhouse iceboxes.

Today, Devil's Lake has stabilized at a maximum depth of about 45 feet, a roughly rectangular lake 1.3 miles long and half a mile wide. When viewed from afar, the uppermost bluffs look much as they did before the glaciers moved in, when the gorge was 300 to 400 feet deeper than it is now. Steep cliffs in any gorge tend to indicate that it was created fairly recently in geologic time, as erosion has not had time to round off the cliffs. For this reason, geographer Lawrence Martin, in his classic book *The Physical Geography of Wisconsin*, described the Devil's Lake gorge as "steep-sided and youthful."[7]

Devil's Lake State Park was established in 1911, but it had long been a destination for people from all around. Native Americans regarded the lake as a sacred place, as reflected in their names for the lake, which translate variously to Spirit Lake, Mystery Lake, Holy Lake, and Lake of the Red Mountain Shadows. It was a place for hunting and possibly for fishing, but it was also a place for wonder and worship of the spirits.

Later, while loggers stripped much of the surrounding forestland, several forested areas of the park were spared and are today considered old-growth forests, according to the park's longtime naturalist Kenneth I. Lange, now retired.[8] Iron miners explored the park but found no sites to mine, although limited amounts of iron mining went on in other parts of the Baraboo Hills. The railroad came to the gorge in 1872, and from that time on, a tourist trade began to flourish.

Before the 1800s ran out, Devil's Lake was the site of four hotels, the most luxurious of which was the Cliff House, built on the north shore in the shadow of the east bluff. Cottages also sprouted, and at one point, as many as nine passenger trains per day were arriving at the lake's railroad station. Excursion boats circled the lake. One side-wheeled steamboat carried up to 100 passengers. A golf course occupied what is now the Quartzite Campground. This flurry of commercial tourism came to an end within three decades, however, and a spirit of preservation drove an effort to have the gorge—indeed, all of the Baraboo Hills—set aside as a park. The short-lived Cliff House was demolished in 1904, and the other three hotels were sold to the state in 1911. None of those buildings remain today.

The end of commercial tourism did nothing to slow the flow of tourists. As car ownership grew in the 1900s, people no longer had to rely on the railroad to get them to the lake, and they came day and night. The iconic chateau building on the north end of the lake was a popular dance hall and night spot in the 1920s and on and off since then.

Being so heavily visited, the new state park seemed like an ideal place to put a Civilian Conservation Corps (CCC) crew to work in 1935. The CCC's dual purpose was to make the cliffs and lake less dangerous and more visitor friendly, while attempting to preserve the natural features that were drawing the tourists. Because of its beauty, the park was considered a plum assignment for the CCC workers. Within six years, they built the camp where they lived (10 or more buildings about where today's group camp is, east of the south shore beach area), the bathhouse and picnic shelters, the parking lots, several other buildings, and the trails that have since given access to the park's magnificent views for millions of visitors.[9]

Devil's Lake is the most popular of all state parks, receiving 1.5 million visitors a year, on average. They hike upon 1.6-billion-year-old quartzite bluffs. They sail, swim, and dive in the 16,000-year-old lake. They camp on the moraines where ancient walls of ice once stood. Because Baraboo quartzite is one of the hardest-known rock types on the planet, rock climbers love the Devil's Lake gorge and have found 2,000 or more routes among the cracks and crags on its walls. Students of all ages and their teachers and professors arrive by the busloads throughout the warmer months of the school year, and scientists from all over the world have probed the park's glades and forests for decades.

For many, including the park's retired naturalist Kenneth I. Lange, it is the stunning awareness of a sense of place that this park brings. In so many spots along its trails, you cannot help but stop and gaze and breathe and wonder.

Lange said it well in a description of one morning in the park, written in the epilogue of his book *Song of Place: A Natural History of the Baraboo Hills*:

> I recall a dawn at Devil's Lake when the air was so calm and bathed in such an orange glow that it seemed as if some great mystery was to unfold. . . . Such magic moments enrich one's life beyond measure. They represent the reward . . . of staying in one place and keeping watch.[10]

TRAIL GUIDE

East Bluff, CCC, Grottos, and West Bluff Trails

These four interconnected trails together provide a grand tour of Devil's Lake and its bluffs and beach areas. The East Bluff trailhead is on the northeast corner of the lake out of a parking area straight east of the park headquarters. It is a moderately difficult trail, with a few steep sections. Much of it is paved with asphalt and includes flights of stone steps built by CCC workers.

A little after one-tenth of a mile, the trail splits, the East Bluff Woods Trail going left and circling through the woods behind the bluff. I took a right onto the East Bluff Trail. After two-tenths of a mile of moderately steep asphalt and stone steps, the trail levels off and turns sharply right, continuing up toward the bluff top. To the left of this turn is a shallow cave tucked in under massive blocks of quartzite. It is called Elephant Cave.

A spur trail branches to the left past the cave. It is an unmarked dead-end trail, but a short distance beyond the cave, it leads to an excellent example of a conglomerate wall made of quartzite rocks and Cambrian sandstone layered in with larger quartzite blocks (Figure 4.6). Geologists interpret this as evidence of an ancient cliff side on an island in the Cambrian sea. [11] The main cliff is farther south, so it is likely that the rounded boulders and stones were loosened and dropped by powerful waves lashing the shore and then rounded by centuries of wave action. Later, they were buried by other sediments and compressed into conglomerate as the sea level rose and more layers were deposited.

Another sign of an ocean shore that you will see on this and other trails is ripple marks frozen into flat rock surfaces. These are from ancient shorelines where waves rippled the sand just as they do today on sandy beaches. As the seas quietly advanced and coastal waters deepened, these ripple marks were buried by muck and the remains of animals and plants that later became layers

4.6 This conglomerate was created around 500 million years ago, at the base of a Cambrian sea cliff. This wall segment is approximately five feet high.

of sedimentary rock. During metamorphosis, such interlayered ripples were preserved in quartzite. Look for them on the flat sides of boulders, in the talus that you will view on some trails, and in some of the stones that now serve as steps on the trails.

Back on the main trail, it is a short climb past the cave to the first view of the lake from on top of the bluff. As you venture onto the bluff top, remember that you stand atop a mass of quartzite that is 4,000 feet deep and about 1,600 million years old.

From here, you begin a rocky up-and-down traverse on a trail across the top of the bluff, passing a distinctive large quartzite boulder called Elephant Rock. Near this rock, at 0.4 mile from the trailhead, is another example of conglomerate created by waves crashing against an ancient cliffside. At several points on this part of the trail, you can walk to the edge of the bluff for views of the north beach area and the moraine behind it. That is where the 200-foot wall of ice stood for about 3,000 years, dropping the gravel and boulders that now underlie the forested ridge. The city of Baraboo is also visible farther north.

After following the bluff for about a half mile, the trail veers away from the bluff and runs through the woods. It turns back out to the bluff closer to the south end of the lake. From the vantage point mentioned at the start of the Devil's Lake section of this chapter, you can find places to sit and take in much of the gorge. This hike description will proceed east on the East Bluff Trail, but from this point, you can also descend to the valley floor using the Balanced Rock Trail. It is one of three trails that descend across the talus field, and it is probably the steepest, most difficult, and most exposed of these trails. However, it affords superb views of the south gorge and leads you to the famous Balanced Rock, sculpted from quartzite by countless cycles of freezing and thawing of water in the crevices of the quartzite mass around it.

From the south end of the lake, the East Bluff Trail turns east and continues along the rim of the gorge with spectacular views of the forested moraine below rising away from the flat south beach area. Here was another wall of ice 15,000 years ago. If you could take yourself back to that time and wait long enough, you might hear and see a block of quartzite peeling away from the rim of the gorge and crashing down onto the talus slope in front of that wall of ice.

About a third of a mile after it turns east, the Devil's Doorway Trail splits off to the right. This is a short loop trail that takes you to the iconic rock formation that is probably the most photographed of all the park's attractions (Figure 4.1). It is a very steep trail with high stone steps to manage, but it is only a tenth of a mile long and provides more great views of the south gorge and quartzite formations.

When you gaze on seemingly precarious formations such as the Devil's Doorway and the Balanced Rock, you might wonder what it would take to tip them over. It would take more than we can easily imagine, because quartzite is extremely dense and heavy rock. Just one cubic foot of quartzite weighs 165 pounds. Try to figure how much the Balanced Rock weighs.

Farther along on the East Bluff Trail, a second route to the valley floor splits off to the right. This is the Potholes Trail, another steep hike across the talus field. The descent to the talus is extremely steep and challenging. However, you need only descend 40 or 50 steps to see the namesake for the trail—the potholes that were drilled into the quartzite by a powerful ancient river that flowed across the area hundreds of millions of years ago. They are near the junction with the East Bluff Trail, on the west side of the Potholes Trail.

Continuing on the East Bluff Trail past that junction, it is another half mile to the CCC Trail, the third and last path down to the valley floor. From this junction you can continue eastward to hike along the bluff, but the East Bluff Trail ends here. What continues east is part of the Devil's Lake Segment of the Ice Age National Scenic Trail. It eventually descends the bluff and crosses the line of the moraine, dropping into Roznos Meadow.

The CCC Trail begins with switchbacks traversing the sheer walls at the top of the bluff. On this trail, you might pass rock climbers with their ropes dangling from the bluff top. This is a popular access trail for climbers. While the hike across the talus below the peak is moderately steep, it is not as challenging as the others. It provides the best unblocked views of the south gorge, along with good closeup looks at the talus (Figure 4.4). It also passes a few flat rocks on which to sit and rest and contemplate the vast gorge and all that has happened here. It is 0.3 mile from top to bottom.

The CCC Trail descends to meet the Grottos Trail, which runs 0.7 mile west along the base of the talus field. An easy, wide trail, it follows a gentle downslope across outwash from the south gorge glacial wall to the flat parking area near the lake's south shore. From this trail, you can step down into the grottos—low spaces off the trail, walled by boulders at the bottom edge of the talus slope. From there, you get a stirring view of the talus, looking up at the enormous rock pile and the cliffs 500 feet above you, from which these boulders have fallen over thousands of years. In such a space, even on the hottest days of the year, hikers enjoy the cool air flowing down through the talus field.

Heading west on the trail, you pass the lower end of the Potholes and Balanced Rock Trails. At the latter junction, the Grottos Trail ends. But a well-traveled path goes from the junction to the south shore parking area and at this point, you have merged with the Devil's Lake Segment of the Ice Age National Scenic Trail. Follow it across the beach area to the boardwalk on the south shore of the lake. From there it is about a mile to the crossing over Messenger Creek and the junction with the Tumbled Rocks Trail, an easy one-mile trail along the shore to the north end of the lake. If you continue a few more yards past this junction, you reach the intersection with the West Bluff Trail.

The West Bluff Trail is a one-mile trail comparable to the East Bluff Trail in difficulty, with several steep sections and stone steps. Much of it is paved in asphalt, but some parts are eroding and require careful stepping. The trailhead is at the junction of Cottage Grove Road and South Shore Road. The first stretch of about a half mile is a steep climb on blacktop and rocks going away from the lake and circling back through woods toward the bluff. At the 0.6-mile mark is the first spur trail to the bluff where the view of the lake and the East Bluff is splendid (Figure 4.7).

From this point, note as you look at the East Bluff that the vegetation pattern slopes down about 20 degrees from south to north. This shows that the rock layers of the bluff are tilted, due to the crumpling of the land described at the beginning of this chapter. The bluff you see is literally a cross-section of the south range of the Baraboo Hills. Recall that the north range was tipped to almost vertical, while the south range was tilted much less steeply, and here you see that tilt clearly.

From the West Bluff, you can also clearly see how both ends of the gorge are blocked by a moraine where walls of ice once stood. In Ice Age days, you would see the East Bluff between the ice walls with no ice on top, but no trees either, only lichens and a few other primitive species. The moraine connecting the two plugs of ice would be out of sight, several miles to the east.

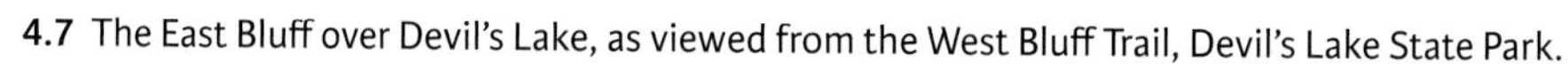

4.7 The East Bluff over Devil's Lake, as viewed from the West Bluff Trail, Devil's Lake State Park.

At about the halfway point on this trail is a developed observation area with benches, a viewing scope, and an emergency call box. The trail continues with some steep stretches and finally skirts the north end of the bluff and switches back to drop more gently down across the north moraine into the north beach area, about a half mile west of the East Bluff trailhead parking area. At this point, you have completed a grand loop that gives you a sampling of all the major geologic features of Devil's Lake and its bluffs.

TRAIL GUIDE

Parfrey's Glen

Beyond the eastern border of the main part of the park is another area of state park land enclosing Parfrey's Glen, a steep-sided hollow created by torrential meltwaters from the glacier. The 0.7-mile out-and-back trail in the gorge is mostly easy with a few short steep sections. As you work your way back into the glen, it becomes narrower and deeper and more other-worldly (Figure 4.8). Beyond the marked trail, you can scramble over a jumble of boulders and ford the stream to get to the end of the glen where the stream cascades down through the hills. Hiking around and beyond that waterfall is prohibited for the sake of the rare and fragile ecosystem there.

On the trail, you encounter outstanding examples of conglomerate containing rounded quartzite boulders. Look for layers of this type of rock high up in the walls of the glen. They indicate the presence of an ancient sea cliff on an island in the Cambrian sea. It was battered by high waves that tore chunks of rock loose from the cliff and tumbled them for centuries at the foot of the cliff to round them off. These rocks and boulders were then buried in sand and gravel and converted to conglomerate layers. According to geologists, the cliff is still there to the west of the trail, but buried under a few hundred feet of sedimentary rock and soil.[12]

In the glen, you will see remnants of a more elaborate hiking trail that once included steps, boardwalks, benches, and viewing platforms that allowed for easy access to the waterfall area. In 2008, these structures were torn up by heavy floodwaters that also dislodged large boulders and uprooted trees. Park managers decided that rebuilding the trail was probably not worth the cost, given that heavy rainstorms can turn the little Parfrey's Glen Creek into a raging torrent.

4.8 Parfrey's Glen on the south flank of the Baraboo Hills was carved by glacial meltwaters thousands of years ago.

Mirror Lake State Park

A few miles north of the Baraboo Hills, Dell Creek flows out of the eastern fringe of the Driftless Area heading southeast toward the Baraboo River. About 20 miles from that river, the stream takes a 90-degree turn to the northeast and flows toward the city of Wisconsin Dells and the Wisconsin River. On this short northeasterly stretch, it flows into a narrow canyon of Cambrian sandstone with 60-foot walls on either side, where it forms Mirror Lake.

This long, snaking body of water is Wisconsin's oldest artificial lake, created in 1860 by the flour miller Horace LaBar, who built a log dam and a flour mill on Dell Creek near the village of Lake Delton. LaBar powered his highly productive mill by controlling the flow of the creek, and the lake he created was originally called LaBar's Pond. The original dam was replaced by today's concrete dam in 1925.

The lake is actually a series of wide sections of Dell Creek, with a long, narrow tributary section that stabs southward into the valley of a smaller creek that flows from the highlands just north of the Baraboo Hills. This tributary section of Mirror Lake is at the heart of the state park, flanked by hiking trails on the west side and a beach and campgrounds to the east. Tourists enjoy the lake for its sandy beach and boating, and the trails are well maintained for summer hiking and cross-country skiing in winter. Mirror Lake State Park was once part of a summer cottage subdivision developed by landowner Bunnel Page. The state acquired the land and opened the park in 1966. It now comprises 2,000 acres and a variety of ecosystems, including bluff-top forests, oak savanna, and small patches of prairie. The park offers a quiet, relatively uncrowded alternative to nearby Devil's Lake State Park.

While the clean white beach sand was trucked in long ago as part of the cottage area development, there is no shortage of sand underlying the beach and every other area of the park. Mirror Lake is on the east edge of the Driftless Area, so it was never buried under ice, but the effects of the glacier made the park what it is, and one of those effects is a deep layer of sand. Another is the lakeside bluffs that give the park its picturesque identity (Figure 4.9).

Before the glacier came, the park area, like all of southern Wisconsin, was probably a less-eroded version of today's Driftless Area. Dell Creek is thought to have flowed generally southeast into the Baraboo River. Then came the damming of the Wisconsin River by the encroaching wall of ice and the flooding of the entire central plain by Glacial Lake Wisconsin. The Mirror Lake State Park area was flooded by part of Lake Baraboo, one of the southern extensions of the lake (Figure 4.3). Over hundreds to thousands of years while the wall of the glacier stood three to four miles to the east, this fjord-like bay draped a layer of sand over the hills and valleys of the park. You do not have to dig down very far to find this sand, and it lies at the surface in many areas.

As the climate warmed and the ice began to melt, large volumes of sand and gravel washed away from the glacier wall. Some of this outwash flowed into the Dell Creek valley and blocked the creek's southeasterly flow, causing it to turn to the northwest, heading for the Wisconsin River. When the ice dams and fragile moraine dams burst and the drainage of the vast Glacial Lake Wisconsin began, Dell Creek became a roaring torrent of ice water that poured through the creek valley, quickly carving it deeper. After a few days or weeks, the lake was gone and what was left was the gorge of Mirror Lake State Park along with other such canyons, including the Dells of the Wisconsin River a few miles to the northeast.

Right after that icy flood, the bluffs that now give the park its character stood slightly higher over the again-docile Dell Creek. Their freshly fractured

4.9 Echo Rock is a popular attraction at Mirror Lake State Park, carved by water during the drainage of Glacial Lake Wisconsin.

Cambrian sandstone faces would have been brighter in color. The bluffs were topped only by a deep layer of wet sandy soil on which today's thick pine and oak forests would develop over the centuries to come.

TRAIL GUIDE

Northwest Trail

This is a 2.3-mile trail rated as difficult by the park. Most of this trail is wide, well maintained, and well marked, but it has a few very steep sections that can be bypassed on alternative routes. In addition to being a good hiking trail, it makes an excellent—although difficult—cross-country skiing trail.

The trailhead is behind the park entrance office across the entrance road. Several cutoff trails allow you to shorten the loop, but by bearing right at the forks on this trail, you take the outermost loop, which gives you a chance to warm up on the flatter first half of the trail before tackling the hillier second half. After one-tenth of a mile, the trail passes the south end of the tributary section of the lake on which the beach is located farther up the shore on the east side of the lake. The trail crosses the stream that feeds this part of the lake on a sturdy wooden bridge and then forks. The right fork runs in a northerly direction along the west side of the lake.

The forest along the trail is typical of the surrounding area. The first part of the trail runs through a forest of mostly maple with a few white pine, which shortly gives way to more white pine and oak as you go farther north. Starting at the 0.8-mile mark, the trail crosses short drainage valleys running steeply down out of highland woods toward the lake.

At close to a mile, the trail passes to the right of the foundation of an old building: the Page family's eight-bedroom summer home. These ruins are steadily being reclaimed by the forest. A little farther on, the trail passes an old cement-and-steel guardrail on the right, separating the trail from the steep decline to the lake. It bordered an old road that served the landowner in days gone by.

The trail next passes through open woods parallel to the bluff next to the lake. It is studded with large old white pines—the closest thing to old-growth forest in the area. The scene harkens to days when white pine forests dominated most of Wisconsin before they were logged off by the early twentieth century. At the 1.2-mile mark, the trail leaves the bluff area and turns south into deeper woods. The sand underfoot becomes more evident here, as red pines now join the mix of trees. They like sandy soil.

About 1.4 miles from the trailhead, you see a high sandstone outcropping off the trail to the left. It looks as if it could be a secondary river bluff lying inland from the creek. In glacial times, when the ice was melting and Dell Creek was wider and deeper, this cliff probably formed the riverbank. The trail wraps around and ascends to the top of this outcropping, then levels off.

After another third of a mile, you enter a much hillier area. There are three points on the trail where you have a choice between a steep section and a more level bypass. This feature was probably designed for cross-country skiers, although it is also helpful for warm-weather hikers. A beginning or intermediate skier or hiker would want to take the bypasses because the hills are very steep. After the two-mile point, the trail levels off and borders what appears to be an old prairie or savanna. It then closes the loop at 2.3 miles.

Rocky Arbor State Park

The dictionary defines *arbor* as "a leafy, shady recess formed by tree branches." "Rocky arbor" is a good description of the intimate, rock-bound, pine-studded acreage that is Rocky Arbor State Park. Another description might be "lost canyon," for that is what it feels like—a place lost on the map and in time.

As the crow flies, Rocky Arbor State Park lies five miles straight north of Mirror Lake State Park and just a mile west of the famous Dells of the Wisconsin River. Like the Dells and Mirror Lake, Rocky Arbor was formed by the torrent of icy water that gushed out of Glacial Lake Wisconsin after its ice dam collapsed thousands of years ago. This little park contains some of the state's best examples of sandstone formations exquisitely carved by moving water.

Before the glacier came, Rocky Arbor was part of the larger eroding sandstone plain that had covered much of the state for many millions of years. When the ice mass rolled in, it stopped just a few miles to the east of the park area, but it flooded the park site by damming the Wisconsin River. The upland area of the park is covered with sand and other lake bottom sediments underlying the maple and pine forest that grows well on sandy soil.

The reason Rocky Arbor feels lost is that it was in fact abandoned by the Wisconsin River after the glacial lake drained and as the glacier retreated. That older version of the Wisconsin River flowed south to where Coldwater Canyon meets the river valley, just as it does now (Figure 4.10). However, instead of branching around Blackhawk Island as it does today, the river flowed only west around the island. Instead of turning southeast at the west end of that island, it flowed southwest into what is now Rocky Arbor State Park.

Within a half mile after exiting the park flowing south, the river jogged southwest again to the point where, today, I-94 crosses the old river valley just north of Exit 89. Just beyond this crossing point, the ancient river did a sweeping left turn (where Exit 89 is today) and headed east along what is now the roadbed for State Highway 13/23 and the creek bed for Hurlburt Creek. That creek flows into the Wisconsin River directly west of the city of Wisconsin Dells. This is where the ancient abandoned riverbed rejoins the Wisconsin River today.

This was the flow that was established sometime after the glacial lake had drained and the area's streams had all stabilized. One such stream was a tributary to the Wisconsin River that flowed west through Artist Glen (Figure 4.10) and then south along the east side of what is now called Blackhawk Island. On the south side of that island (not an island at the time) was a short tributary flowing east to join the Artist Glen tributary. As with many streams, its headwaters were gradually migrating back, eroding the underlying sandstone westward toward the old Wisconsin River channel. When it reached that channel, the eroding stream opened a gap in the south bank of the river, and the river's flow gradually shifted to the tributary streambed.

In *The Physical Geography of Wisconsin*, Lawrence Martin interprets the change of course of the Wisconsin River as a case of stream capture, also referred to as stream piracy, a process by which a new channel for a stream opens up and captures (becomes the path of least resistance for) the stream's water, diverting it from its former route.[13] In this case, the diverted river's volume and flow were considerably larger and faster than those of the tributary stream that captured it and the river bored a wider and deeper streambed fairly quickly. The new riverbed was thus established, and the old bed where Rocky Arbor lies became a shallow wetland in a canyon about 1.5 miles to the west of today's Wisconsin River.

This state park is considered a quiet alternative to the far more bustling Wisconsin Dells area. Its sights are not quite as spectacular as those of the Dells, but it has its own form of ancient, majestic beauty. Its bluffs are 90 feet above the valley floor, so it hosts upland and lowland vegetation, wet and dry plant communities, and a diverse collection of birds and other animals. Rather than

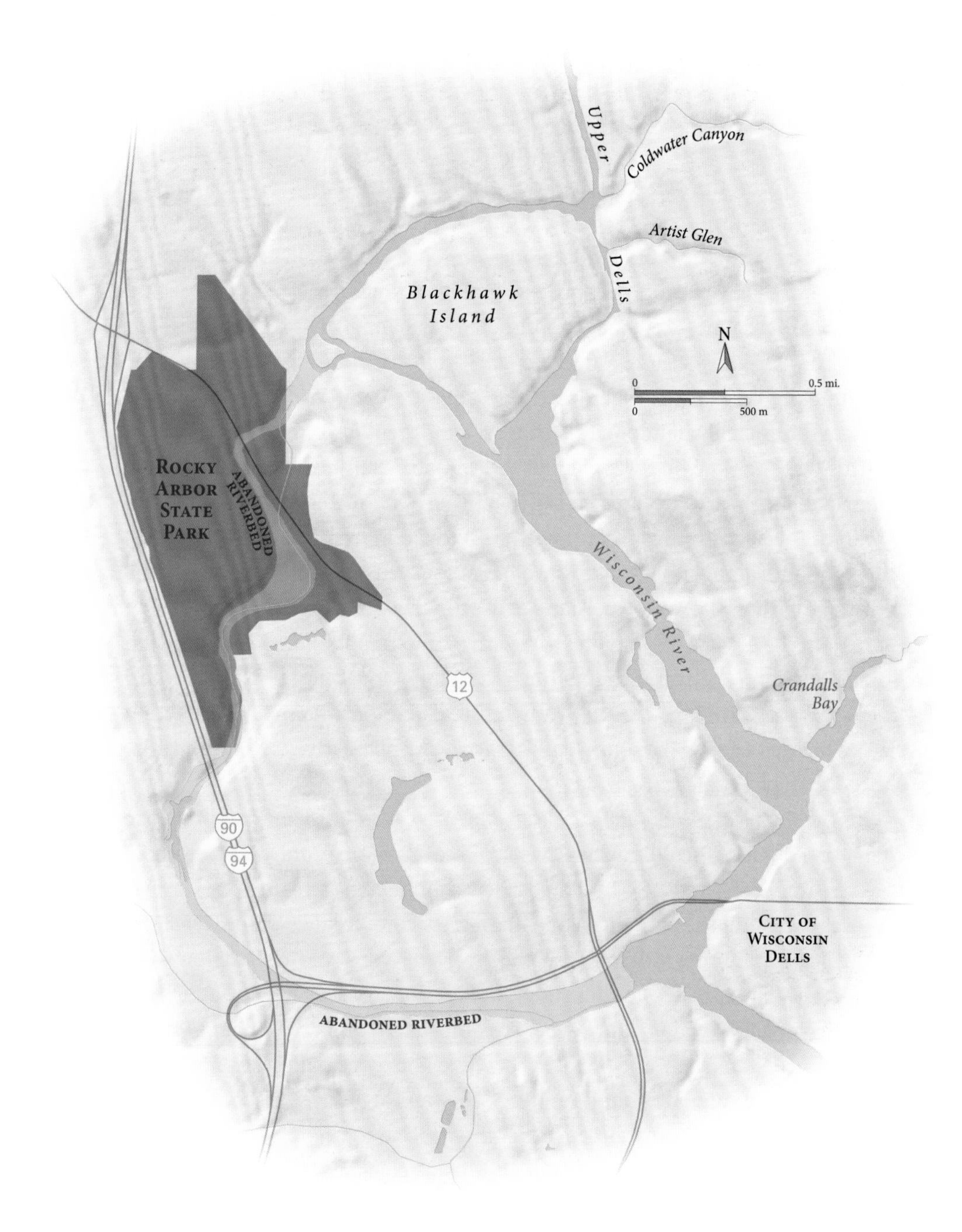

4.10 The ancient Wisconsin River abandoned its course through Rocky Arbor State Park and now flows in a channel to the east of the park. MAPPING SPECIALISTS, LTD., FITCHBURG, WI

relying on a boat to take you near the sandstone formations as you do in the Dells, you can walk right up to them in Rocky Arbor. Some of the water-sculpted formations are amazingly intricate and beautiful (Figures 4.11 and 4.12).

In the early decades of the twentieth century, the park area was slated for highway development. The state had purchased the land for that reason, but to the relief of those who valued its quiet beauty, the state government canceled the highway plans and turned the land over to the Department of Natural Resources. The 250-acre park was established in 1932.

4.11 (*above*) The sandstone bluff at Rocky Arbor State Park was carved by moving water when the Wisconsin River flowed through this canyon.

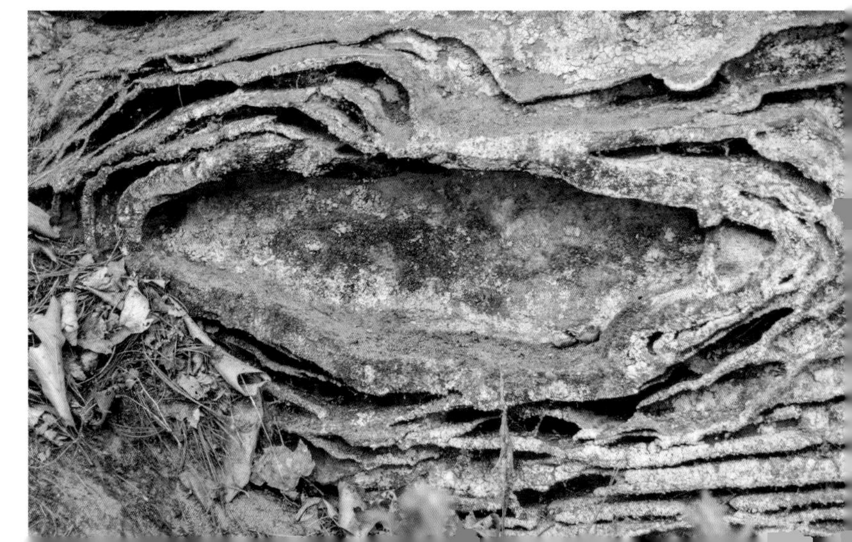

4.12 In this closeup view, the intricate effects of weathering and flowing water on the sandstone are visible.

TRAIL GUIDE

Rocky Arbor Trail

The trail into the old gorge is an easy two-mile loop. It runs between the wetland at the bottom of the gorge and the base of the pine-studded sandstone bluffs on the west side of the gorge. It then ascends the bluff, runs along the top for a short distance, and goes back down into the gorge.

The trailhead is at the southwest end of the picnic area parking lot, not far from the park entrance, in a fragrant grove of red and white pines mixed with maple trees. The first stretch of the trail takes you to some superb examples of water-carved sandstone formations (Figure 4.11), many of them now being overgrown by the forest. At the half-mile mark you pass an often-photographed lone pillar of sandstone on the left. It was a little island in the old river.

As you hike, look up at the bluffs and notice huge blocks of layered sandstone tipping away from the bluff as if they will fall any day. You are watching the super-slow-motion process by which a mass of sandstone breaks apart due to wind, water, and frost erosion. This along with the prying roots of trees and other vegetation are gradually dismantling the bluffs.

At the 0.75-mile mark, a staircase takes you up to your right over one of the sandstone outcroppings. At the top of the staircase, the trail ascends to the top of the bluff, with some moderately steep segments. It then curves to the north, running through a pretty maple forest on top of the bluff, behind the rock outcroppings you saw below.

After a mile of hiking, you come to a place where you can look to your right for a view down into the old river gorge. If it were about 10,000 years ago, shortly after the retreat of the glacier, you would see the ancient Wisconsin River flowing there. The walls of the gorge would be freshly carved sandstone in bright shades of tan and brown, bearing no vegetation. The flowing water would be washing against the bases of the bluffs, patiently working the sand grains loose and carrying them downstream.

In another quarter mile, you pass a spur trail that cuts to the left toward the park's campground. The main trail continues back down through outcroppings toward the trailhead. At the 1.5-mile mark is another wooden staircase going to the floor of the canyon, and from these stairs, you get another good view of the old gorge. In another half mile, you are back at the trailhead.

Mill Bluff State Park

The western shore of Glacial Lake Wisconsin was ragged, running in and out of the deep hollows of the Driftless Area (Figure 4.3). Mill Bluff State Park sits near that ancient shore and is located in the Driftless, but its character is defined by the events of glacial times.

The park contains several steep-sided sandstone structures, including mounds, buttes, and pinnacles, standing as high as 200 feet above the surrounding flat, sandy plain formed by the glacial lake between 19,000 and 14,000 years ago (Figure 4.13). They are standing fragments of the Cambrian sandstone mantle that once covered most of Wisconsin before being worn away by millions of years of erosion.[14]

Why, you might wonder, did these sandstone structures survive all that erosion? Geologists think that in the places where they formed, water was seeping from the ground for millions of years, bringing calcium carbonate to the sandy surface. The sandstone bodies that were formed within these pools of calcium carbonate solution were much harder than other types of sandstone around and underneath them.[15] When the surrounding sandstone dissolved away over the millennia, these harder sandstone bodies acted as umbrellas, shielding the softer sandstone under them from the elements. They are often referred to as caps protecting the softer sandstone towers and pillars beneath them.

When the glacial floodwaters rolled in, the lake depths in the Mill Bluff State Park area are thought to have reached 60 to 80 feet, and the taller sandstone buttes became islands in the icy lake. Wave action around these islands took over the erosive work that wind and rain had done for millions of years. It continued picking apart the softer sandstone under the hard sandstone caps while lake currents distributed the newly eroded sand across the lake bottom. When the softer sandstone in a butte was carved deeply enough from under its cap, the weight of the capstone would cause pieces of it to fracture and plunge into the lake. When all of a cap was taken down in this way, it was just a matter of time until the underlying structure became a mound of dissolving sandstone and eventually disappeared. It seems reasonable to conclude that there must have been many more sandstone islands in the lake when it first formed, and that they have since been dismantled.

Throughout the lake's existence and especially when the glacier started melting, there were icebergs breaking away from the ice wall and floating in the lake.

Many of them carried boulders the glacier had picked up, and as the icebergs melted, these boulders became glacial erratics scattered on the ancient lake bed. Some of these icebergs rammed into the sandstone islands and got stuck there. In several cases, boulders on these ice rafts also became lodged on the sides of the islands where the icebergs had rammed them, and there they remain today. Geologists have used these erratic markers to estimate the depth of the glacial lake and how it varied over time.

The glacial lake drained away and the sandy, silty lake bottom was eventually reclaimed by dry land and wetland plant communities. Ancestors of the Ho-Chunk established communities across south-central Wisconsin. They and other Native American travelers used the buttes and pinnacles of the Mill Bluff area as landmarks to guide them on the trails, as did later European explorers and settlers.

Because the buttes and other sandstone formations are more delicate than they look, one person climbing or carving on them can do as much damage as decades or centuries of erosion can do. For this reason, preservationists have

4.13 Sandstone buttes in Mill Bluff State Park as viewed from atop Mill Bluff. The tallest bluffs used to be high, rocky islands in Glacial Lake Wisconsin.

managed to have the Mill Bluff area structures protected, first as a state park, which was established in 1936. In 1971, Mill Bluff got additional protection when it was made a unit of the Ice Age National Scientific Reserve. While the Civilian Conservation Corps was active in many parks, it was the Works Progress Administration, another federally funded program of the 1930s, whose employees worked at Mill Bluff State Park. Among other projects, they installed the 223 stone steps that now take visitors to the top of Mill Bluff.

TRAIL GUIDE
Nature Trail and Bluff Trail

Combining these two trails into one hike—viewing the bluff from below and then climbing to the top—will give you a full appreciation of Mill Bluff and similar sandstone buttes. Except for the climb, this is an easy, gently rolling trail. The climb to the top on stone steps is steep in places, but the steps are solid, as is the iron railing next to the steps. At the top, the trail is level and easy and less than 0.2 mile, running the full length of the bluff top.

The trailhead for the Nature Trail starts at the easternmost parking area, farthest into the park from the entrance. The trail begins by going north along the east side of the bluff. Through the woods you get good views of the bluff top to your left, rising 120 feet above the plain. Near the north end of the bluff, large sandstone boulders lie on either side of the trail. They appear to be a harder form of sandstone and probably were parts of the resistant capstone that broke off sometime during the past 20,000 years and tumbled to the old lake bed.

As you look up at the top of the bluff, note that its upper reaches are vertical or very steep. These walls of sandstone descend to a more gently sloping skirtlike formation draped around the entire bluff. This is the deep pile of sand dissolved from the softer sandstone beneath the hard cap on top of the bluff. Eventually, when the cap disintegrates, the butte will dissolve to a mound of sand that will in turn melt into the plain over time.

One of the interpretive signs on this trail informs us that on the east side of the bluff, conditions are cool and moist because it is sheltered by the towering wall of stone from westerly winds and the hot afternoon sun that have dried the soil on the west side of the bluff. Consequently, this side supports a more diverse mix of trees and other vegetation, including oaks, maples, and ferns. As you round the north end of the bluff about a half mile from the trailhead and proceed south along its west side, note that the forest there is dominated more by white and red pines, which grow more easily in sandy soil under warm, dry conditions.

At 0.6 mile from the trailhead, the Nature Trail meets the Bluff Trail, which is a long flight of stone steps ascending to this junction from the trailhead on Funnel Road across from the beach area. The steps, installed by the Works Progress Administration, start out on a gentle slope and get very steep as you approach the top of the bluff. The heavy iron railing is dependable—a must for any hiker's safety.

As you climb, note the sandstone outcroppings to your right—a striking example of layered Cambrian sandstone deposited 500 million years ago. As you climb, every foot of stone you pass represents hundreds of thousands of years of deposition by ancient rivers flowing to a sea. Note the layer of fine sand eroded from the stone on some steps, reminding you that erosion goes on.

The trail on top of the bluff runs north-northwest to south-southeast. At the northwest end is a viewing platform with stunning views of the old lake bed stretching to the north and east, studded with the buttes, mounds, and pinnacles that remain as remnants of the ancient sandstone mantle. As shown in Figure 4.13, the closest butte is called Bee Bluff. It is 60 feet tall, so probably was under water during much of the lake's existence. (In this area, recall, the lake was 60 to 80 feet deep.) The pair of mounds centered behind Bee Bluff is called Camel Bluff, which stands 170 feet tall. The structure visible between and behind the humps is Devil's Monument. The more distant mound at far right is Ragged Rock, which has lost its protective cap and is therefore eroding more quickly than the other buttes in this photo. The park's largest butte, Long Bluff, standing 199 feet above the plain, is not visible from this viewpoint.

If you take the short bluff-top trail to its south end, you can look west into the Driftless Area and see more mounds, buttes, and pinnacles south of the park. Standing atop Mill Bluff, picture this: all the land between you and the distant bluffs was once a rolling sandstone plain that has since dropped away due to timeless erosion. If you wonder where all the sand in this part of the state came from, imagine all of that sandstone for 30 miles to the north and east being dissolved, and you will have part of the answer.

After you descend the steps to the junction with the Nature Trail, it is a short hike down the slope to the parking area. The Nature Trail loop, without the climb to the top, is just under a mile long.

TRAIL GUIDE
Camel Bluff Trail

This is a gentle, easy, and well-maintained loop trail, 1.25 miles long. The trailhead is on the north side of Interstate 90-94 off Funnel Road, which runs between Mill Bluff and the Camel Bluff Trail.

The sand that underlies this whole part of the state is evident in many places on the trail. The entire area has just a thin covering of topsoil and vegetation over this deep body of sand. It reminds you that you are strolling on the ancient lake bed where sand collected for 3,000 or more years. The forest that took hold here is dominated by maples, oaks, jack pine, and white pine.

The trail starts by going north along the west flank of Camel Bluff's northern hump. Camel Bluff is actually a pair of sandstone bluffs connected by a low-rising saddle formation. Shortly after turning east and skirting the north end of the north bluff (0.6 mile from the trailhead), the trail curves southeast and runs through a remnant oak savanna prairie. There you can look west and see Devil's Monument, Cleopatra's Needle, and the north end of the bluff rising from the plain. It was probably under water when the lake was at its deepest.

The trail then angles southwest and passes a classic example of a pinnacle, or pillar of rock (Figure 4.14), of which there are several on the sand plain. A spur trail to this formation allows a closeup view and clear evidence of how wave action has worked on this rock spire.

About a mile from the trailhead, the trail rises gently and runs up and over the low saddle between the two humps of Camel Bluff. To your left is a bold rock outcropping that forms the northern tip of the south bluff. If the glacial lake were still here, you would be under water and parts of Camel Bluff would be standing close to 100 feet above the lake's surface. At 0.2 mile beyond the saddle, the trail closes its loop and turns back toward the parking area.

Roche-a-Cri State Park

The geological story of Roche-a-Cri State Park is much the same as that of Mill Bluff. The towering butte named Roche-a-Cri is a remnant mound of Cambrian sandstone that was long protected by a cap of harder rock (Figure 4.15). At 300 feet tall, it stands more than twice as high as Mill Bluff above the surrounding

4.14 One of several pinnacles of sandstone on the ancient bed of Glacial Lake Wisconsin in Mill Bluff State Park.

plain. It is also more isolated, standing alone and farther from any edge of the ancient glacial lake bed (Figure 4.2). For much of the lake's existence, it must have been an impressive island, rising 150 feet or higher above the lake's surface.

Almost all the literature on this park translates *roche-a-cri* as "rock with crevice." The thought has been that the French explorer who came up with the name was referring to a large cleft in the top of the mound. However, the name literally translates to "rock that cries" and the Frenchman was more likely referring to the cries of eagles and hawks that frequent the crags of the bluff top.[16]

Unlike Mill Bluff, the hard sandstone cap that protected the softer sandstone in the body of this bluff is gone, and Roche-a-Cri is now eroding faster than in centuries past. Compared to Mill Bluff and other buttes closer to the Driftless Area, Roche-a-Cri stood in deeper water after the glacial lake invaded and suffered more powerful wave erosion. Currently, the top layer is about 20 feet of sandstone remaining from a layer deposited late in the Cambrian period. Seas of that time hosted primitive life, including wormlike creatures that lived in clusters of vertical tubes on the sea floor. Geologists have found fossil remains of these tube clusters in the top layer of sandstone on Roche-a-Cri.

4.15 Roche-a-Cri stands alone, 300 feet above the surrounding plain.

This park provides ample evidence of human occupation in the distant past, including the most well-preserved Native American rock art in the state. It can be viewed from a fully accessible platform at the south end of the butte. These carvings indicate that Roche-a-Cri was a special, perhaps sacred place for early Native Americans who used the nearby stream for transportation. Ancestors of today's Ho-Chunk tribe are thought to be the creators of this rock art.

European settlers tried to farm the sandy plain surrounding Roche-a-Cri. Because the soil drains so quickly, they soon found that farming was difficult to do without almost constant irrigation. Farmers also found that the wetland areas of the plain, where the water table is high, are suitable for growing cranberries, and this is now a leading crop in the area. The view from the top of the bluff includes broad areas covered by commercial cranberry bogs.

With the European immigrant settlements came changes to the area's ecosystems. Before the land was farmed, fire swept the area's prairie and savanna lands regularly and vegetation had evolved to live with it. For example, white oaks grew well in the sandy soil and tolerated fire and thus oak savanna came to dominate many areas of the sand plain. European settlers claimed the land, cleared much of it, and protected it from fire, and soon white pine seedlings took hold where before they were regularly destroyed by fire. The white pines grew quickly and shaded out the oak acorns, which need full sun to germinate.

White oak is still prominent in many areas of the sand plain, but it is being replaced by white pine and other species. Since presettlement days, forests have replaced large areas of prairie and savanna. Birds that had lived there, including meadowlarks, grouse, and ring-necked pheasant, as well as the animals that hunted them, moved on as the forests moved in.

Roche-a-Cri has been owned by the state of Wisconsin since 1937 when it was acquired by the State Highway Commission as roadside property along State Highway 13. It soon became a recreational area and the Civilian Conservation Corps did some work at the future park. The Conservation Commission (an earlier version of the Department of Natural Resources, or DNR) established Roche-a-Cri State Park in 1948. The DNR put up the protective viewing platform and interpretive panels near the rock art in the early 1990s, and a wooden stairway to the top of the mound had been built by 1994.

Today, visitors climb the mound on sturdy steel staircases mounted on wooden posts. This ambitious project to replace the old wooden stairway involved the use of a helicopter to set the heavy new staircases in place to be secured by workers on the bluff. Most of this project was completed one day in 2012, and the helicopter work was challenging and potentially dangerous because winds can be unpredictable around high, rocky cliffs. According to park manager Heather Wolf, it was one of the most nerve-wracking days she ever spent at the park.[17]

One of the ongoing projects in the park is prairie restoration south of the mound (see Trail Guide). Roche-a-Cri Mound has been set aside as a state natural area, which gives the rock art and fragile sandstone cliffs further protection against overuse and abuse by park visitors. An interpretive sign in the park describes it well: "One of 300 Natural Areas that preserve native plant and animal communities for education, research, and the benefit of future generations."

TRAIL GUIDE

Acorn and Mound Trails

This is an out-and-back hike on a combination of trails. The park includes more than six miles of trails that meander around the mound and across the ancient lake bed. All trails are gentle, wide, and well maintained. The only steep section is the 300-foot climb on the stairway to the top of the mound.

A good place to start is at the Winter/Prairie Parking Area, so named because it borders the restored prairie and provides parking for cross-country skiers during winter. You can start your hike by studying an impressively thorough interpretive sign that tells all about the native prairie plants being reestablished in the park. The first half mile is a hike on level land through the prairie, and the south end of Roche-a-Cri is visible the whole way (Figure 4.15). The trail comes to a T where it meets the Acorn Trail, a 3.5-mile loop trail. Take a right at the T and a quick left at the junction of the Acorn and Spring Peeper Trails to stay on the Acorn Trail. A half mile from the parking lot, you pass a pretty picnic area next to a stream very near the rock art viewing platform. Here is where Native American rock artists may have landed their canoes to go to work on the rock.

The rock art is in two forms—petroglyphs, or rock carvings, and pictographs, or rock paintings. They are thought to have been created by ancestors of the Ho-Chunk, the petroglyphs around 1,100 years ago and the pictographs 400 to 500 years ago. No one knows exactly what they represent, but interpretations include clan symbols, stories of visions experienced by the artists, mythical bird/man ancestors, and depictions of the hawks, eagles, and vultures

4.16 The Roche-a-Cri mound represents many millions of years of sandstone deposition.

that soar around the craggy peak of Roche-a-Cri. The crescent shapes among the carvings might represent astronomical events such as solar or lunar eclipses. Other markings on the rock include the initials and dates carved much later by travelers, including European settlers and American cavalrymen.

The trail continues north around the mound from the rock art viewing area. One of the remarkable features in this park is the fact that you can hike right to the base of the mound and get a closeup look at its sandstone layers (Figure 4.16). This is one of the best views you can find of such ancient sandstone beds, and as you gaze straight up at the towering butte, you are looking at many millions of years of sandstone deposits.

The Acorn Trail is aptly named, for you can feel and hear the acorns crunching underfoot as you stroll through the oak and pine forest around the mound. At 1.1 miles from the parking area, the Mound Trail splits off to the left, while the Acorn Trail continues to the right. The Mound Trail runs to a picnic area and playground on the north side of the mound. Across the playground is the staircase to the top of the bluff.

It is well worth the challenge to take the stairs to the top, and there are several benches built into the stairway for resting. Along the top of the mound is a similarly built walkway, designed to keep visitors off the bluff top itself. Even a few people per day hiking and climbing directly on the bluff top would quickly accelerate its erosion and the demise of the bluff, so this is an important improvement in the park.

The overlook platform on the north end of the walkway is sturdy and broad, affording spectacular views of the ancient lake bed to the west, north, and northeast. Park managers have mounted informative signs that identify the many mounds and buttes that remain on the ancient lake bed. They appear in every direction with names like Sheep Pasture Bluff, Duckworth Ridge, and Rattlesnake Mound. Take some time to enjoy this view and to remember that 200 million years ago, this was a level plain lying on the level at which you stand.

After descending to the base of the bluff, you can rejoin the Acorn Trail for a three-mile meander through the forest and around the mound area. It gives you a good sampling of the oak-pine dry forest that dominates this part of Wisconsin's Central Sand Plain.

5.1 Dundee Mountain in the Kettle Moraine State Forest is one of many outstanding features created by the glacier.

5

The Glacial Showcase

Southeastern Wisconsin

Most of the southeastern quarter of Wisconsin lies flat and rolling—a docile landscape, ideal for farming and for building towns and cities (Figure 5.2). It was not always so. Stepping back in time, we see events and processes that formed a bedrock in southeastern Wisconsin that belies the gentle, rolling topography.

About 1,600 million years ago, the continental collision that built the Baraboo Hills forged other small mountain ranges across the southern half of the state. Within several hundred million years, these ranges were buried in sediment from Precambrian seas. Using drilling and other research techniques, geologists have detected two of these ranges—one southwest of Lake Winnebago, the other northeast of Waterloo in Dodge County. The tallest of these quartzite peaks, the buried cousins of the Baraboo Range, are thought to have stood 700 feet or more above the surrounding Precambrian bedrock. Some scientists speculate that over time, erosion will expose them as it has the Baraboo Hills.[1]

Of the seas that encroached and receded several times throughout history, the Silurian sea is of particular importance to eastern Wisconsin. It covered the entire state for periods between 444 and 417 million years ago. Its waters over eastern Wisconsin were warm, shallow, and clear—ideal for the growth of coral reefs, underwater structures built by lime-secreting algae and tiny animals called polyps. Some of the Earth's earliest known coral reefs grew and thrived in the Michigan Basin. While the western margin of these reefs reached only as far as the Milwaukee area, the Silurian sea hosted a variety of other ancient

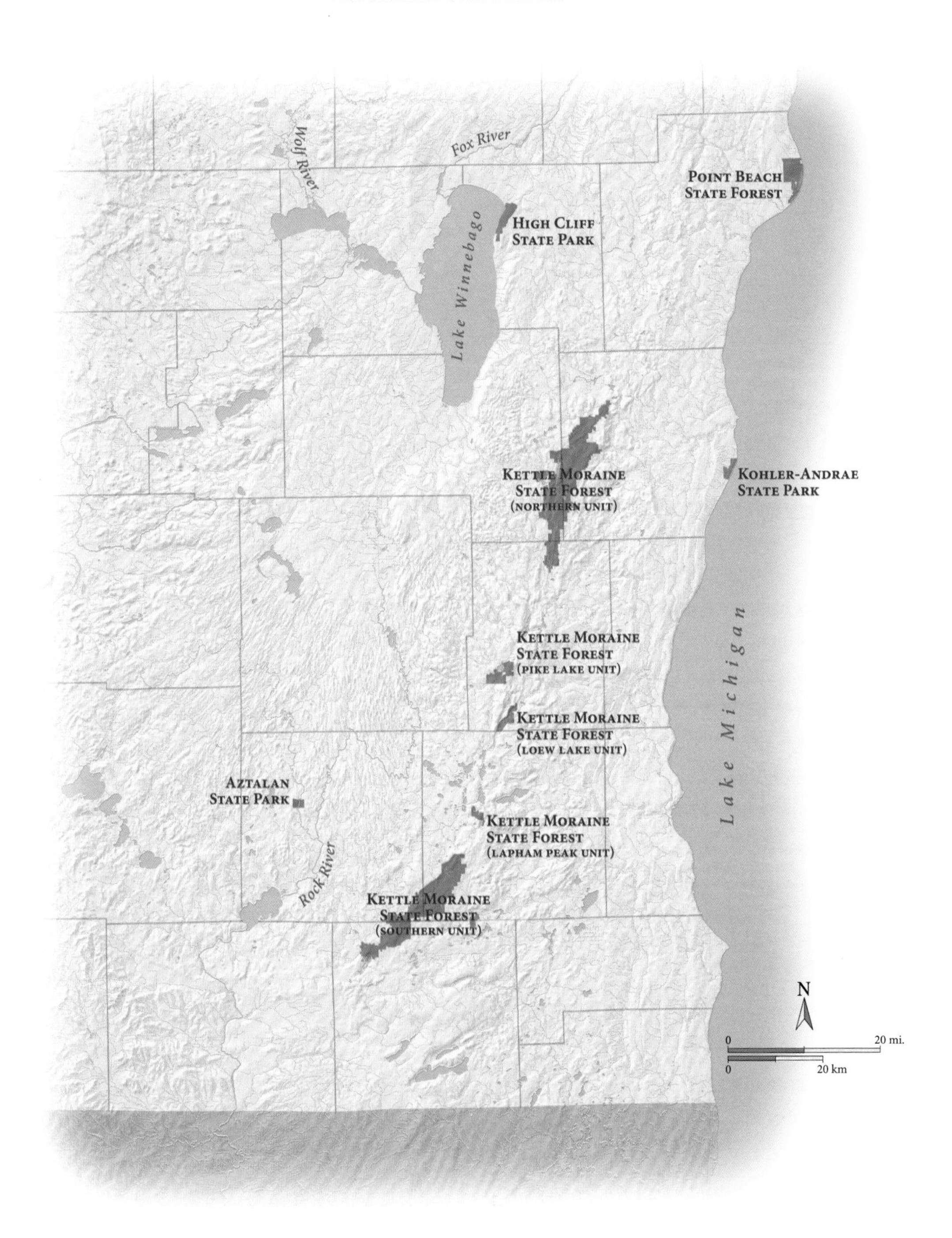

5.2 Southeastern Wisconsin. MAPPING SPECIALISTS, LTD., FITCHBURG, WI

marine species. As these populations thrived and died, their remains collected on the ocean bottom and eventually formed the thick layer of dolomite that lay over all of Wisconsin.

For more than 300 million years after the Paleozoic seas finally receded, wind and water scoured and carved the sedimentary plains that had been ancient sea bottoms. Erosion was the primary shaper of the land, but the underlying bedrock also played a role. On any landscape, erosion goes to work first on the highest points of the uppermost layers of rock, and to the east of the Wisconsin Dome and Arch, the layers of bedrock are sloped from the dome and arch all the way down into the center of Lake Michigan. After erosion took the upper layers off the tops of the dome and arch, the tipped-up edges of the underlying rock layers sloping to the east were exposed. The edges of the softer layers then eroded faster than the edges of the harder layers, leading to a pattern of north-south oriented ridges and valleys across eastern Wisconsin.

In such a pattern, the ridges are made of harder rock such as dolomite. The eroded edge of a tilted rock layer forms a steep side, called an escarpment, and the intact rock layer sloping gently away from that edge forms the other side of the ridge, called the dip slope (Figure 5.3). These asymmetrical ridges are called cuestas. The most prominent cuesta in eastern Wisconsin is the Niagara Cuesta, also called the Niagara Escarpment, so named because it is the same cuesta that forms Niagara Falls on the New York–Canadian border. This cuesta formed just west of Lake Michigan where the sloping land is, as geologists say, dipping into the lake basin. It thus forms the rim of the Michigan Basin, the east side of which includes the escarpment under Niagara Falls. (This is described in more detail in Chapter 6.)

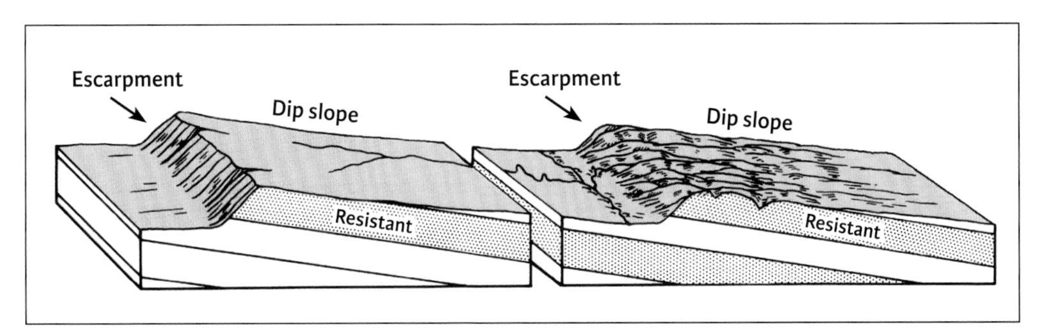

5.3 Cuestas are formed by the erosion of sloping land where the rock layers differ in their hardness.
TREWARTHA, *ELEMENTS OF GEOGRAPHY*. REPRINTED BY PERMISSION OF MCGRAW-HILL

After the long period of erosion throughout late Paleozoic and Mesozoic times, and just before the Ice Age, southeastern Wisconsin probably resembled today's Driftless Area to the west. It would have hosted flat upland prairies, high ridges, and deep, forested valleys. The Niagara Cuesta was a long, linear highland separating lowlands now occupied by Green Bay and Lake Winnebago to the west and by Lake Michigan to the east. These lowlands did not contain large bodies of water in pre–Ice Age days, but major rivers were likely flowing northeast in both valleys, draining the surrounding highlands.

Over the course of about 2 million years, several glacial advances each helped to dig these lowlands a little lower. Of the great lobes of ice that flowed as parts of the Wisconsin glaciation (Figure 1.4), the two that covered the eastern part of the state were the Green Bay and the Lake Michigan lobes, named for the lowlands that channeled their flows. In these lowlands, the glacier was a bulldozer. Most forms of sandstone, in layers from a few to hundreds of feet thick, were easily pulverized and moved by the ice. The glacier literally plowed this rock away, forming the broad basins of Green Bay, Lake Winnebago, and Lake Michigan and heaping the debris mostly at the margins of the basins.

These moving masses of ice picked up all loose and easily loosened material, from silt and sand to boulders 20 feet in diameter. Most of this material migrated up into the ice mass as it flowed along, and much of it rose to the top of the glacier and rode along on its upper surface. This upward movement of material occurred whenever ice being pushed from the glacier's interior rode up over slower-moving ice at the base of the ice mass. It would occur when the ice mass was pushed up a slope or when the ice at the base became frozen to the bedrock. The forward ice would be blocked by the slope or immobile base ice, while the ice in back and on top of it kept moving. Think of waves cresting as they approach a shoreline where surface water keeps flowing and water beneath it is slowed by contact with the shore. The front margin of an advancing glacier is a super-slow-motion version of such cresting.

As some of the ice rolled up over the ice below it, it carried great volumes of sand, gravel, and boulders picked up at the base of the glacier, and this was how this debris moved to the top of the ice mass. There it accumulated in layers and formed something of a landscape atop the glacier. Varying thicknesses of debris on top of the ice, called supraglacial debris, caused different rates of melting. Thick layers of debris insulated the ice underneath and kept it from melting, while exposed areas of ice melted more quickly in warmer temperatures. Thus icy peaks, ridges, crevices, streams, and lakes formed on top of the glacier.

As the ice eventually melted out from beneath the supraglacial topography, it left a newly formed landscape now on display in the glacial showcase that is southeastern Wisconsin. Glaciologists from around the world have come to study these features (Figure 5.4). They include:

- Till: the jumbled assortment of sand, gravel, and boulders that had ridden with the ice sheet, now draped across the land in layers up to hundreds of feet thick.

- Moraines: the ridges of debris built wherever the front edge of the ice mass stood in one place for years or centuries. There the marginal ice would be melting while the forward movement of ice and debris from the glacier's interior acted as a vast conveyor belt, dropping tons of debris to form long ridges of jumbled rock, gravel, and sand. The moraines lying along the farthest advance line of the glacial lobes are called terminal moraines.

- Drumlins: elongated, teardrop-shaped hills, usually occurring in groups called swarms. Drumlins range in height from a few feet to 150 feet above the surrounding plain and average a quarter mile wide by one to two miles long.[2] The steeper side of a drumlin (bottom of the teardrop) faced the ice as it moved over, shaping that steeper front slope and the gentler downstream slope. Some drumlins include sections of bedrock. Others are composed of mostly sand and gravel. The temperatures on the bed of the glacier varied; some stretches were more frozen and much harder than others. Some geologists think drumlins formed wherever the glacier's bed had frozen spots that remained as high ground while surrounding areas were relatively thawed and easier for the ice to plow away.[3]

- Ice-walled lake plains: flat-topped hills that once formed the beds of supraglacial lakes. Streams atop the ice mass carried silt and sand to these ice-walled lakes where the sediments accumulated and leveled the lake bottoms. When the glacier melted, the ice walls slumped away, leaving the lake bottoms as the flat tops of gently sloping hills, typically a mile or more in diameter. Today, farmers drive their tractors upslope to raise crops on many of these ancient lake bottoms.

- Kames (or moulin kames): cone-shaped hills formed of gravel and sand that washed from the top of the glacier down through a vertical shaft called a moulin. These shafts opened as part of the melting process, and supraglacial meltwater streams plummeted through them, creating caverns at the base of the ice mass where debris accumulated. Over centuries, the debris piled up in the telltale cone shapes within the ice caverns, and they remain as distinctive hills today.
- Eskers: long, winding ridges that vary in height from a few feet to more than 150 feet and in width from a few yards to half a mile. They can stretch for miles, although many are just hundreds of feet long. During the melting phase, streams formed and flowed under glaciers. These subglacial streams bored tunnels under the ice, and there they deposited the silt, sand, gravel, and rocks they were carrying. Over centuries, the tunnel walls molded the deposited material into the sinuous ridge shapes that remained when the ice melted away.
- Tunnel channels: troughs cut into the land running in the direction of a glacier's advance, up to half a mile wide and miles in length. When a subglacial stream approached the front edge of an ice mass, it flowed over increasingly less frozen land and thus could cut into and erode

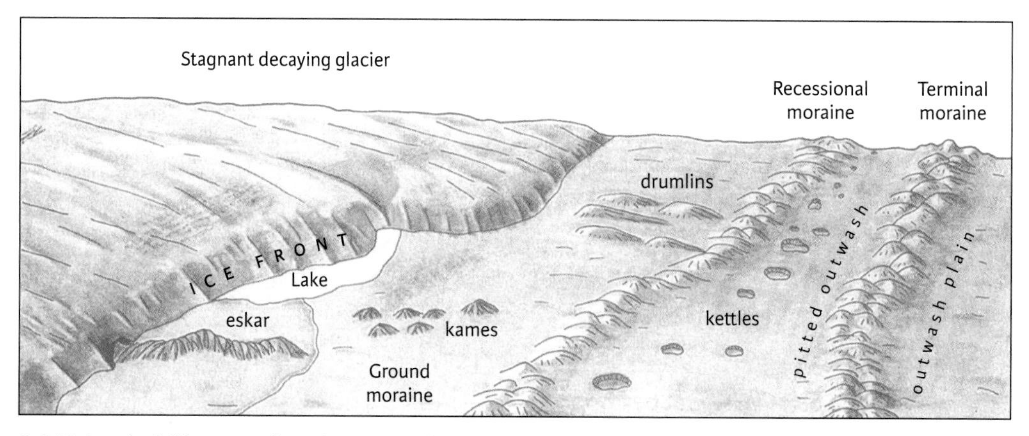

5.4 Major glacial features of southeastern Wisconsin. SCHULTZ, *WISCONSIN'S FOUNDATIONS*. REPRINTED BY PERMISSION OF THE UNIVERSITY OF WISCONSIN PRESS

the underlying bed of the glacier. In Wisconsin, along the line of the farthest advance of the ice lobes, dozens of such streams cut through the terminal moraine, carving broad troughs across the land. Within some of these troughs, the debris-laden streams formed eskers that now run along the bottoms of some tunnel channels. When melting accelerated, the flows in some of these tunnels were enormous, pushing large boulders to the mouths of the tunnels where the water escaped the ice.[4]

- Outwash plains: broad, gently sloping layers of glacial debris spread across the land. When subglacial streams burst from beneath a glacial wall, they quickly spread out and joined other such streams in broad, braided flows of icy meltwater carrying debris onto the ice-free land. In so doing, they lost their velocity and power and thus the largest boulders pushed by these streams stopped rolling just outside the mouths of the ice tunnels. However, the flow was still powerful enough to carry gravel, sand, and silt away from the ice mass to points where the water could no longer move them. In this way, the streams sorted the glacial debris, with boulders, gravel, sand, silt, and clay particles landing in that order in areas from nearest to farthest from the ice wall. Thus an outwash plain is a sorted and layered apron-shaped formation.

- Kettles: depressions in the Earth scattered across wide areas, typically bordering the edges of a glacial ice mass, where chunks of ice separated from the glacier melted slowly as the glacier retreated. As meltwaters flowed, much of this lingering ice was covered by debris and remained buried for centuries after the main body of the glacier had retreated. Before the buried ice masses melted into the thawing ground and flowed out in springs, they represented highlands and domes. After the complete melting, these higher areas collapsed into the spaces once occupied by the ice chunks. Many of the resulting depressions filled with water and became kettle lakes, and this is how more than two-thirds of Wisconsin's 15,000 lakes were formed.[5] Others remained wet but were filled with vegetation, becoming kettle wetlands and peat bogs. Still others dried out and became deep wooded hollows. Kettles can be a few dozen feet in diameter or several miles across and a few to more than 100 feet deep.

- Hummocky terrain: glaciated areas where hills and ridges are randomly interspersed with kettle lakes, wetlands, and ravines. Like kettles, this terrain typically exists in areas near the glacial moraines where masses of stagnant ice lingered for years or centuries and were buried by outwash and wind-blown sediment. As the ice melted out, the highlands that had been on top of the ice masses dropped to form kettles. The areas between kettles that had been lowlands became ridges and hills, called hummocks. Glaciologists call this process topographic reversal. In areas where debris had lain thickly on top of the ice mass, such as in northern Wisconsin, the hummocky terrain is more dramatic. In southeast Wisconsin, the terrain is of lower relief because the ice was not as rich in debris.

- Glacial lakes: bodies of meltwater that collected in areas where the glacier wall was retreating. These lakes filled basins and were contained by highlands on all sides except where they sat against the ice wall. (The process of glacial lake formation is described in more detail in Chapter 4.) Lake Michigan was formed in this way and Lake Winnebago is a remnant of such a lake. Over centuries, glacial lakes deposited sediments on their floors in layers that amounted to more than 100 feet thick in some places. Today those nutrient-rich sediments help to nourish crops grown on large areas of the ancient lake bottoms.

The paths of the Green Bay and Lake Michigan lobes largely determined the topography of southeastern Wisconsin. The Green Bay lobe flowed into the lowland just west of the Niagara Cuesta now occupied by Green Bay, Lake Winnebago, Horicon Marsh, and the upper valley of the Rock River. It pushed up and over parts of the Niagara Cuesta, trimming much of it off in some areas.[6] The Lake Michigan lobe shaped the lake basin and where it met the Green Bay lobe, the two masses of ice covered a large area with a double moraine, now called the Kettle Moraine.

The most recent glacier entered what is now Wisconsin about 26,000 years ago and took 8,000 years to reach its maximum extent, where it sat more or less in that position for another 3,000 years—until about 15,000 years ago. It then began its halting retreat as the climate began to warm. After a few re-advances, it finally melted away in northernmost Wisconsin around 10,000 years ago. Note that the time from when the glacier first entered the state to the time when it finally retreated was 16,000 years—many times longer than all of recorded human history.

The ice sheet over Wisconsin was unimaginably massive—far deeper at its center than any sea that had ever covered the area. Scientists think it was around 10,000 feet thick at its deepest near the center of the mass, with 1,000 to 2,000 feet being the more typical thickness in most areas. It tapered to hundreds of feet thick near its advancing edge, where it stood as a long, steep wall of ice. So massive was the Laurentide Ice Sheet in Wisconsin that it compressed the land. The thicker the ice, the greater the compression, so the land was warped downward to the north and east, compressed by 300 feet or more in northern and eastern Wisconsin.[7] Since the ice retreated north, the land has been rebounding slowly and still is today, at an estimated rate of about one foot every 100 years.[8]

Shortly after the ice was gone, the land had little or no vegetation. Southeastern Wisconsin was a rolling plain, covered by a wet mix of boulders, gravel, and sand. On the barren, slippery slopes of the higher moraines, drumlins, and kames, landslides were frequent as erratic boulders that had been dropped there by the glaciers loosened and rolled down rocky hillsides, taking trains of gravel and sand with them.

As the ice retreated farther north, the changes in vegetation and animal life were the reverse of the changes that had taken place as the ice moved in. Gradually, areas of tundra south and west of the ice sheet were taken over by spruce forests. The animals of the tundra, especially frigid-weather mammals such as the wooly mammoth and musk ox, followed the glacial mass north. The spruce forests were favored by browsing species, such as the American mastodon and white-tailed deer, and their predators including wolves and large cats. Humans arrived around 12,000 years ago as new predators of the big-game animals. Spruce forests slowly gave way to red and white pine and then to deciduous species such as oaks and maples. The earliest natives of eastern Wisconsin continued to hunt large and small game and began cultivating wild crops to supplement the plants gathered from the prairies and forests for food and medicine.

Except for some of the mammals that went extinct shortly after the Ice Age ended, the flora and fauna of eastern Wisconsin came to resemble those of presettlement times. They are in turn reflected today in areas where presettlement prairie and forest ecosystems have been restored.

The glacial land features found here represent a wealth of resources in the forms of gravel, sand, and stone used for constructing roads, fences, and buildings in cities and on farms. With the intensive development that began and accelerated after Europeans arrived in the Wisconsin region, people mined these resources heavily, erasing many of the glacial features. Drumlins, eskers, and kames were damaged or destroyed, literally hauled away to provide these building materials. In the late 1950s, concerned citizens organized an effort to preserve some of these features, resulting in the establishment of nine Ice Age National Scientific Reserve areas, some of which overlap with the parks and forests covered in this book. Another such endeavor resulted in the Ice Age National Scenic Trail, which is maintained by a remarkable group of volunteers. They are part of the Ice Age Trail Alliance, which also works on land protection and management and provides education on Wisconsin's glacial legacy, much of which is now available for all to study and enjoy.[9]

Aztalan State Park

On a late summer afternoon, Aztalan State Park, when viewed from its entrance, can seem a ghostly place. The land slopes away to the east, and spread across it are long rows of tall posts, arrayed as walls, and broad earthen pyramids. On a clear evening when these structures are lit sharply by slanting sunlight, they are reminiscent of the ruins of ancient Aztec and Mayan cities found in Mexico and Guatemala. They can give a visitor the feeling of having strayed far to the south and back in time.

In the 1830s, American explorers of the site had similar reactions. One of them, Nathaniel Hyer from the nearby village of Milwaukee, came to map the site and study its mysteries. He borrowed from an earlier explorer and writer, Alexander von Humboldt, who had surmised that the Aztecs had arrived in Mexico "from the North; from a country which they called Aztalan."[10] Hyer took the liberty of assigning that name to this mysterious ancient village in southeastern Wisconsin.[11]

Since then, archaeologists working extensively at Aztalan have found that the people who built the village were actually not connected to the Aztecs of Mexico, but had arrived there from a much closer location—an ancient town on the Mississippi River near present-day East St. Louis, Illinois. They were part of a widespread civilization called Mississippians. During its peak, the Mississippian influence had spread throughout the Mississippi River watershed from the upper Midwest to the Gulf of Mexico and from the Atlantic coast to the Great

Plains in the west. Its capital of 10,000 to 15,000 people was called Cahokia, from which the Aztalan settlers are thought to have come in about 1100 CE. (Cahokia may have supported a much larger population of at least 35,000 in the region around the city.) Cahokia, Aztalan, and other Mississippian settlements shared common social structures that were reflected in well-organized towns and cities and distinctive architecture, including the earthen platform mounds now found at Aztalan State Park.[12]

However, archaeologists have also determined that the Mississippians from Cahokia were not the only occupants of Aztalan. They were joined by, or they joined, a group of Late Woodland people who were living in southern Wisconsin at the time the village was built. We know this from the mixed remains of pottery and other items found at the site that clearly came from both groups and were left there at the same time.

The question of why Cahokians ventured north to this site has several possible answers. Some archaeologists think it was a breakaway group that moved north to establish a new chiefdom, having had disputes with the power structure in Cahokia. Others suggest that the expanding population of Cahokia created a need for other settlements. The Mississippians were active traders, so perhaps the northern community was intended to be an extension of the trading network or an outpost from which the trade routes could be protected.

Whatever the reason, it is interesting to imagine the first group of scouts venturing north and finding the site of Aztalan. Perhaps they paddled up the Mississippi to where the Rock River flows into it, near present-day Rock Island, Illinois. They may have explored tributaries of the Rock River, including the Crawfish River and, following that stream, found the sloping expanse of prairie land that now hosts Aztalan State Park.

The site would have been highly attractive, both for its resources and for its strategic access to waterways. On the west side of the Crawfish River where the town was to be built was a sloping prairie with small stands of trees, suitable for clearing and settling. The top of that slope overlooked the river, which was bordered on the far side by a mixed-hardwood forest. It would have hosted abundant edible plants and populations of white-tailed deer, other smaller mammals, and game birds. These woods would also provide timber for firewood and building material.

During the time when Aztalan was built, the Crawfish River is thought to have widened just downstream of the site to form a large, shallow lake with

5.5 An artist's conception of the ancient walled village of Aztalan as viewed from the east, across the Crawfish River. WHI IMAGE ID 28935

abundant fish and clam populations and wild rice beds. These waters were frequented by ducks and other waterfowl. The river's floodplain was fertile and well drained, ideal for raising the corn and squash crops commonly grown by the Mississippians. Wherever the multiple freshwater springs fed the river, they prevented it from freezing during most winters, which would allow a supply of freshwater year-round, or nearly so.

All of this added up to an extremely attractive site for a new Mississippian town. In fact, it had long been such an ideal site. Archaeologists have found evidence of previous smaller and less permanent Native communities occupying the site long before Aztalan, taking advantage of the verdant prairie, forest, and river features that had been molded by the glacier. These glacial gifts provided resources for the mysterious culture, remnants of which are now preserved within Aztalan State Park.

The Mississippians brought something new to the region around Aztalan—an organized town structure reflected in elaborate buildings and walls intended to be permanent (Figure 5.5). They enclosed 21 acres inside walls built of tall poles woven together with willow branches and covered with fire-hardened clay. They must have been much more imposing than the skeletal representations that have been reconstructed in the park. These walls included 32 evenly spaced watchtowers. The builders also constructed three large earthen structures—pyramids composed of platforms (Figure 5.6)—on three of the town's four corners.

5.6 A reconstruction of the largest of the three platform mounds that existed in Aztalan.

Within the walls, the town was divided by similarly massive walls into three zones: a residential area for commoners stretching along the riverbank, a public plaza in the area just upslope, and in the higher side of the town, a residential and ritual area reserved for the chiefs and other elites of the community. In their book *Aztalan: Mysteries of an Ancient Indian Town*, Robert A. Birmingham and Lynne G. Goldstein note: "This water-to-sky arrangement, with intermediating ceremonial grounds and people, modeled the Mississippian universe."[13]

Here is a short summary of what researchers have inferred from archaeological evidence.[14] The population of Aztalan grew to as many as 350 people, all living within a strict social structure. The common citizens provided food and hides for clothing by hunting, fishing, and farming. Farmers raised mostly corn, but also squash, pumpkins, and sunflowers. Hunters brought home deer and game birds, while fishers provided fish and clams. Others gathered berries and nuts from wild plants. Food was collected and stored in several large pits in the town, to be distributed during ceremonial feasts and other rituals. Commoners' houses, spread along the lower section of the town, were typically small, each with just enough room for a food storage pit, a fire pit, and places to sleep.

Trade was an important part of the Mississippian culture. Based on the presence of certain types of pottery, pendants, and stone hoes and other tools, archaeologists surmise that Aztalan was well connected to Cahokia and its trade network. Those who made stone tools and implements used a certain kind of sandstone from west-central Wisconsin, lead from southwest Wisconsin, pipestone from Minnesota, and copper from Michigan.

Warfare was evidently a common part of Mississippian life. One indication of this is the use of high, massive walls and watchtowers around the towns. Human skeletal remains found at Aztalan indicate that violent deaths were common. Violence was a theme in Mississippian art, as well. Early scholars thought this evidence also suggested that residents of Aztalan engaged in cannibalism, but later studies have found this to be highly unlikely. The question of who the Mississippians were fighting is not well understood and is a subject of ongoing research.

Aztalan was occupied for about 150 years and abandoned by around 1250. Within another 150 years, Cahokia had also been abandoned, and the Mississippians vanished from the Midwest, possibly moving to what is now the southeastern United States. Possible reasons for this mysterious relocation are disease, depletion of forest and prairie resources including soil and firewood, and political upheaval due to ongoing warfare.

Evidence indicates that another group of Native Americans, probably the Ho-Chunk, lived in and near the Aztalan site in the late eighteenth or early nineteenth century. They established camps and left items such as iron knives, fish hooks, and musket parts, but their communities were nowhere as robust as was Aztalan 500 years earlier.

In the 1830s, European settlers coming to the Aztalan site found long ridges of clay and dirt, mixed with charred remains of human-built structures. Word spread quickly and relic hunters soon arrived in growing numbers to poke through the ruins of Aztalan. They dug into the mounds and excavated randomly in search of antique treasures. One of the more respectful explorers was the renowned naturalist Increase Lapham, who took a scientific approach to exploring Aztalan, making the first detailed map of the site in the 1850s. It was published in his landmark book, *The Antiquities of Wisconsin*.

In the late 1800s, uncontrolled relic hunting continued at Aztalan while farmers also moved into the area to clear and plow the land. Farmers dug up and hauled away the clay and charred remnants of the walls, and some even tried to plow down the ancient platform mounds at the site to make more space for planting. By 1900, the entire site of Aztalan was planted in corn. Many of the features described on Lapham's map had been destroyed.

In 1919, anthropologist Samuel A. Barrett, working for the Milwaukee Public Museum, began directing a careful scientific exploration of Aztalan. His staff spent two years on an archaeological dig at the site—the first such exploration anywhere in Wisconsin. Barrett continued his research and in 1933 published *Ancient Aztalan*, a seminal volume that spurred later research and captured public attention. Barrett had painted Aztalan as an archaeological treasure worthy of preserving. By then, the Wisconsin Archaeological Society had purchased a small part of the site and, working with the State Historical Society of Wisconsin and the Lake Mills-Aztalan Historical Society, had become involved in educating the public and bringing pressure to bear on the state. Their efforts led the state Department of Conservation (now the Department of Natural Resources, or DNR) to purchase the entire site in 1948 for development of a state park, which was officially established in 1952.

In 1964, the US Department of the Interior listed Aztalan as a National Historic Landmark. Even so, the park has been threatened more than once by lack of funding and subsequent decay and vandalism. In the 1980s, the people of Aztalan Township stepped up to maintain the park and keep it open. Today, the Friends of Aztalan State Park, formed in 1994, play such a supporting role in preserving the treasure that is Aztalan. Research also continues on the site. In addition to archaeological digging, researchers now use far less invasive remote sensing to detect buried pits, posts, and artifacts. As a part of public education efforts, they have reconstructed platform mounds and lines of posts like those that once formed the skeletons of Aztalan's massive walls.

TRAIL GUIDE

Ancient Village Trail

From the park entrance off County Highway Q, the park road takes you to the far southeast corner where you can start your tour of the grounds of the ancient village. From the parking area, walk east toward the river to the mowed, grassy trail that loops the perimeter of the site. The trail includes numerous signs giving information about the building of the town, its daily life, and archaeological studies of the site.

In the southeast corner, you enter what was the residential area of the town, stretching north along the river. This strip of riverside land was where most of the daily living took place among the humble dwellings of the villagers. The area was completely enclosed by high walls. The low mound you pass as you begin walking north was a cemetery of sorts where many bones were found by early explorers. At the north end of the residential area is the reconstructed northeast platform mound that probably formed the base of a temple and meetinghouse for the commoners of the village. Archaeologists note that such buildings on mounds were typical in Mississippian towns and that people kept sacred fires burning continuously within such shelters.

The grassy path then approaches the north end of the town and turns left, or west, to cross the plaza area. In the days of ancient Aztalan, you would have had to pass through a guarded gate in the high wall. Within the plaza—probably a largely open, flat, dirt-covered area surrounded by high walls—the people held their feasts and other ceremonies and played games. Large storage pits at the north end held food and other supplies throughout the winters. Researchers think the dirt from the many storage pits scattered within the town's walls was used to build the platform mounds.

At the northwest corner of the town, on the other side of the plaza wall, was another platform mound. If you were a commoner in ancient Aztalan, you probably would not be allowed to cross this wall. The mound was within the part of town controlled by the town's chief and his family or clan and perhaps one or more spiritual leaders. This elite area enclosed the plaza on three sides and was protected by high walls on all sides. This northwest mound was thought to be the site of a mortuary building dedicated to the chieftains and their relatives.

Beyond the outer wall of the town, farther up the hill to the northwest and just north of the park entrance, nine conical mounds remain. They are called the Ceremonial Post Mound Group, because archaeologists determined that each of them formed the foundation for a huge post nearly two feet in diameter at the base. No one knows their exact purpose, but the posts were thought to have been important for some ceremonies, possibly including burials. Lapham's map showed dozens of such mounds running from the southwest corner of town to the northwest corner and beyond. All but the ones you see here were leveled and plowed under.

From the northwest mound, turn south and walk alongside one of the reconstructed rows of posts that formed the framework for a wall. You are approaching the largest, most impressive mound of the three (Figure 5.6). This was where the chieftains' residences were located and where sacred rituals involving the chieftains and other elites of the town took place. Just west of this mound was the highest point in the town. Lapham's map shows that it was once enclosed by a separate wall extending from the main wall on its southwest corner. The purpose of this extra enclosure remains a mystery.

You can ascend the large southwest mound on steps set in as part of its reconstruction. You might choose to sit for a while on one of the steps to imagine life in Aztalan more than 800 years ago. Sit still long enough, and you might almost hear the sounds of the people in the plaza and in the riverside residential area down the hill from you as they go about their daily activities—hauling water or a catch of clams and fish up from the river, working in the crop fields beyond the walls, toiling to maintain the massive walls, preparing meals, or playing games in the plaza. It was a place starkly different from what it is now and from what it was before the Mississippians came to spend their 150 years here.

Kettle Moraine State Forest

When the glacier moved into eastern Wisconsin, the Green Bay and Lake Michigan lobes flowed generally south, but also pushed southeast and southwest from their centers (Figure 1.4). The Green Bay lobe overrode part of the Niagara Escarpment and pushed farther east to where it met the Lake Michigan lobe flowing west and southwest in what is called the interlobate zone. Both ice masses sloped down from their centers to their margins, so on the surface of the ice lobes where they met was a long, southwest-trending low area. There the gigantic conveyor belts in both lobes piled a double load of glacial debris within this long supraglacial valley.

This accumulation took place over thousands of years. When temperatures finally rose, growing meltwater streams began to flow on either side of the long ridge of debris. Meanwhile, this ridge acted as an insulator that protected ice below it from the heat of the sun, so that it melted more slowly than the exposed areas on either side. On the northern two-thirds of the interlobate zone, the streams began to build their own debris ridges, which grew on either side of the central ridge over hundreds of years. In most of the southern third of the interlobate zone, however, while the central ridge collected debris, this double-ridge system did not form.

The glacier retreated between 18,000 and 10,000 years ago, the ice gradually melting out from under these debris ridges. As they settled onto the bedrock, they formed massive moraines. In the northern interlobate zone, the double ridge that had formed on top of the ice became two parallel moraines with a lowland between them. The original central ridge had largely disintegrated, slumping toward the streams on either side of it, as meltwater hauled much of its sand and gravel down through moulins to form cone-shaped piles under the ice. These piles are now kames, several of which sit in the lowland between the two moraines. In the southern interlobate zone, the single debris ridge became a single massive moraine.

The glacier left a barren, cold, and wet landscape with a wide assortment of moraines, kames, eskers, and hummocky terrain. The last features to form were the kettles where glacial debris slowly collapsed into the large pits as they were vacated by buried ice chunks that took probably hundreds of years to melt. They became kettle lakes, bogs, and hollows scattered randomly across the hummocky land on and around the moraines. Outwash plains also formed where meltwater had gushed away from the ice mass, carrying tons of sand and gravel onto lowlands.

This wild and complex assortment of topographical features had been dropped onto the vast area of relatively flat or rolling land that is southeastern Wisconsin. Much of it is now preserved as the Kettle Moraine State Forest, composed of five separate units—the Northern, Southern, Pike Lake, Loew Lake, and Lapham Peak Units (Figure 5.2).

The interlobate moraines and other features of the area are unusual in that they are made mostly of sand and gravel—among the biggest sand and gravel piles in the world—while most such glacial features are composed of till, which is more rocky.[15] This is because they were built more by streams carrying sand and gravel than by slowly moving ice masses that typically dropped everything from clay particles to massive boulders. It is interesting to note that the high ridges of the Kettle Moraine were once riverbeds. In addition to the sand and gravel, the area has plenty of boulders, but they are mostly erratics transported by the glaciers from as far north as Canada.

The Kettle Moraine stretches for more than 120 miles between Kewaunee and Walworth Counties. It varies in width from one to ten miles, and sits 100 to 300 feet above the surrounding land. Although it is a complex arrangement of ridges and hills, the northern part is generally arranged on two parallel ridges. To the south is the single ridge, part of it piled on top of the Niagara Cuesta, which had been overtopped by the glacier in some areas.

This state forest contains some of the most outstanding examples of glacial features in the world. For example, the moulin kames of the Kettle Moraine

(Figure 5.1) are extraordinarily large and "unique in size and number," according to geologist David M. Mickelson.[16] Several outstanding examples of this feature can be found in the lowland between the two ridges of the Northern Unit of the forest. Eskers, hummocky terrain, and of course kettles and moraines are also dramatically displayed in the Kettle Moraine.

The 3,500-acre Scuppernong Marsh, a wetland-prairie complex in the Southern Unit of the forest, once covered tens of thousands of acres in and beyond the state forest boundaries. Another once-dominant plant community that barely exists today were oak openings—small stands of mostly oak trees scattered on the prairies. For centuries, various Native American groups of the area used controlled burning to maintain the prairies as a source of food and medicinal plants. When farmers took over the land, they suppressed the annual fires that had occurred naturally and used the land for grazing and crops. Consequently, the oak openings all but disappeared and were replaced by farm fields and thicker forest stands.

The state forest was established in 1937 and dedicated to public recreation and forestry, but it was never a contiguous area. The first two sections to be set aside were the Northern and Southern Units. Before the state forest was established, more than 50 percent of the Northern Unit was farmed at one time or another, as well as possibly 70 percent of the Southern Unit. Because the units are not contiguous spreads of land, some farming continues on small areas of private land interspersed among state forest lands.

The Northern and Southern Units each have a headquarters building and a nature center. The nature centers are excellent sources of information, thoroughly and colorfully presented, on the geological, natural, and human histories of the area.

Since establishment, the state forest has been expanded. The Northern Unit sprawls across 30,000 acres, the Southern Unit occupies 21,000 acres, and the Lapham Peak, Pike Lake, and Loew Lake Units cover a combined area of nearly 3,000 acres. Together, they contain more than 300 miles of hiking trails, including 60 miles of the Ice Age National Scenic Trail.

Over the years, managers at the Kettle Moraine have shifted from a focus on forestry and recreation to preservation of the geological, ecological, and human historical heritage of the area. This includes prairie restoration projects, educational exhibits in the nature centers and on several trails, and historical sites. For example, in the Southern Unit, DNR workers are trying to restore the Scuppernong Marsh system with the goal of making it the largest lowland prairie east of the Mississippi River.

Historic sites include Old World Wisconsin, which features historical buildings, implements, and information on farming in the nineteenth and twentieth centuries. Located within the state forest boundaries, it is owned and operated by the Wisconsin Historical Society as the largest outdoor museum of rural life in the United States. This living history exhibit employs interpreters in period dress to help visitors appreciate the history on display.

TRAIL GUIDE
Lapham Peak

Lapham Peak, which is close to the Southern Unit and a few miles south of Delafield, is the highest point in Waukesha County, at 1,233 feet above sea level. The hill is named for Increase Lapham, Wisconsin's first scientist, who lived in Oconomowoc and made weather observations from the peak. In addition to his work documenting Aztalan, Indian mounds, and the geography of the entire state, as well as many other pursuits, Lapham was the father of the National Weather Service and made the first-ever weather forecast in the country in 1870. The 45-foot observation tower atop Lapham Peak provides views of kames, eskers, meltwater channels, and other excellent examples of glacially created landforms for miles around.

5.7 Another view of Dundee Mountain, the Kettle Moraine State Forest, Northern Unit.

TRAIL GUIDE
Dundee Mountain Summit Trail

This is a trail to one of the largest moulin kames in the Northern Unit of the Kettle Moraine, called Dundee Mountain (Figures 5.1 and 5.7). Spiraling to the top, the trail is moderately steep but not rugged. You begin this trail from the campground in the Long Lake Recreation Area, north of Dundee. The trail leaves the campground opposite campsite 945. (You can also reach this trailhead from the boat landing, where there is ample parking. A nice trail runs south out of that lot along the west flank of a ridge that borders Long Lake. It is a little over a mile to the trailhead from the boat landing.)

At the 0.2-mile mark, you pass a trail going to your left, one end of the Summit Trail loop. A short distance beyond this junction, the next trail going left is the other end of the Summit loop. On that trail, you first ascend some steps and a tenth of a mile farther on, pass a broad, green outwash plain on your right where 13,000 years ago or so, a braided meltwater stream flowed away from the shrinking wall of ice.

Continue on the trail for about another fifth of a mile, and after a short switchback you come to a vantage point with a wide, clear view to the west. A little to your left is the town of Dundee, and Long Lake stretches away to the north. The highland that you see on the far side of the lake is the western flank of the Northern Unit's double ridge formation described earlier in this chapter. Farther west is a series of clearly defined humps on the horizon. These are some of the famous Campbellsport drumlins, a swarm of hundreds of drumlins that have been studied by glaciologists from around the world.

Just past the half-mile point is the peak of Dundee Mountain (Figure 5.8). There you can rest on a bench and imagine yourself sitting under a torrent of ice water pouring down from the shaft overhead. You would be pelted with sand and gravel brought down by the shower of meltwater. It must have taken hundreds or thousands of years for the falling water to build this peak. Just past this point is a trail sign explaining the ridge that runs away from the peak of the hill. This is a good example of how an esker can be connected to a kame. It makes sense, for once the water in a moulin drops onto the mound of debris beneath it, it has to go somewhere. Many kames have a similar tail of high land stretching away from them where the stream flowed off the peak and continued through a tunnel carved under the ice, forming an esker.

5.8 The summit of Dundee Mountain in the Kettle Moraine State Forest, Northern Unit.

At about the 0.8-mile mark, after a steep descent on hundreds of wooden steps, the trail crosses another grassy outwash plain spreading off to the right away from the kame. You are hiking through a mix of hardwoods, including oak and maple, along with birch and basswood. On the other side of this outwash plain, about another tenth of a mile along, the trail passes a pretty little kettle wetland off the base of the kame. It may once have been a kettle lake that over hundreds of years was filled in with wetland vegetation. From here, you make a short steep climb on gravel, and close the loop on the Summit Trail.

TRAIL GUIDE
Parnell Esker Trail

The Parnell Segment of the Ice Age National Scenic Trail is about 14 miles long, but you can see one of the largest and best examples of an esker by taking a 3-mile segment starting at the Butler Lake trailhead off Butler Lake Road northwest of Dundee. At the trailhead, you see a cross-section of the esker where it was excavated long ago to make way for a lake access road. It is 30 to 40 feet high here and a little wider at its base.

This section of trail, on the eastern side of the Northern Unit's double ridge formation, offers a rugged hike, traversing a few steep, rocky sections where erosion has taken away topsoil revealing the rocks, stones, and gravel brought here by the glacier.

The trail heads north-northeast along the crest of the Parnell Esker on the west side of Butler Lake, a small and peaceful lake that sits within a large kettle peat bog. After about a third of a mile, the trail drops off the esker and crosses a small footbridge over a stream that flows out of the lake. The Parnell Esker is actually a broken string of esker segments because the subglacial stream that formed it did not deposit its load evenly; the esker was more massive in some sections than in others. Although a 14-mile segment of the Ice Age Trail is named after this esker, its total length is just a little more than 2 miles.[17]

The trail skirts the bog extending from the lake and parallels Butler Lake Road near its junction with County Highway V before crossing Highway V at about the 0.9-mile mark. From there, it ascends to a high ridge that parallels the Parnell Esker. At 1.1 miles, you begin passing between two low wetlands. To your left, across the broad wetland, you can see the esker running north-northeast,

5.9 This kettle pond is one of several that can be seen along the Parnell Esker segment of the Ice Age Trail.

parallel to the trail. Here you get a sense for how big an esker can be and for how long it took for the esker to be built, grain by grain, by the subglacial stream flowing under the ice mass.

Along this stretch of the trail, if you look back to your left, you can get a clear view of a large kame—one of several undisturbed kames that lie in the lowland between the two ridges of the Northern Unit. Here, thousands of years ago, ice lay over the area between the two main ridges. As the glacier retreated, several moulins opened and icy water plummeted through them, pulling debris from the older central ridge and dropping it onto the cone-shaped mounds that you see now.

The trail now follows a long traverse on the high ridge between the wetlands. In the spring, hikers are treated to the sounds of woodpeckers squawking, geese honking, and chorus frogs singing for their mates. At the 1.5-mile mark, the trail passes a remarkably thick spruce forest on the right. Taking a side trip into this stand of spruce, you can get a feel for the type of forest that dominated much of the area in the days after the glacier retreated and the land cover slowly evolved from lichens and mosses to forests. Stand there long enough and you might even sense the ghost of a mastodon ambling among the trees.

At the 2.5-mile mark, the trail crosses Scenic Drive and continues on the high ridge passing kettle ponds and wetlands (Figure 5.9). At the 3-mile mark, the trail skirts a little kettle lake and bog on the left. At this point you can either retrace your steps back to Butler Lake or continue on for about two miles to the Parnell Tower, an observation tower affording great views of kames, eskers, kettles, and other features in this part of the Kettle Moraine.

TRAIL GUIDE
Bald Bluff/Stone Elephant Trail

The Blue Spring Lake Segment of the Ice Age Trail, found within the Southern Unit of the Kettle Moraine, is quite rugged, with steep, rocky sections. It starts near Bald Bluff, a 200-foot-high kame that is topped with an oak opening, which was once the dominant ecosystem in this area but is now relegated to just a few places. For that reason, Bald Bluff is a state natural area.

Begin the hike from the Bald Bluff Scenic Overlook parking area on the east side of County Highway H south of Palmyra. After a 0.2-mile moderately steep climb, the parking lot spur trail meets the Ice Age Trail at the point of a

5.10 The Stone Elephant rock formation on the Ice Age Trail through the Kettle Moraine State Forest, Southern Unit.

switchback on its ascent of the bluff. From there the trail reaches the bluff top and the oak opening after another 200 yards. Here is a bench and an open view of the land to the west and southwest. The trail continues east across the top of the bluff about 500 yards before descending steeply down the back side.

This trail provides an excellent sampling of high-relief hummocky terrain, composed of high, irregular ridges, deep kettle ravines, and kettle wetlands, all formed 10,000 years ago when masses of gravel, sand, and boulders sitting atop the Wisconsin glacier dropped to the ground as the ice melted. The trail passes over or around all of these landforms, now covered by a forest dominated by oaks and pines.

About a half mile from the top of the bluff, the trail enters another oak opening and crosses it to enter a stand of white pines. Shortly beyond this stand, after crossing a horse trail, continue straight ahead, keeping an eye out for the yellow blazes that mark the Ice Age Trail. Just a quarter mile beyond this junction, the trail lies along a low ridgetop and then makes a short, steep climb up a well-defined kame.

After dropping down off the little kame, the trail runs for about 500 yards along the top of a well-defined esker that must have been formed by the stream that flowed under the ice and away from the kame. After leaving the esker, the trail crosses more classic high-relief hummocky terrain for a little over 0.3 mile before a spur trail branches to the left, marked by a sign designating the Stone Elephant. This is an impressive granite erratic (Figure 5.10), about 6 feet high and 12 feet long, dropped by the glacier after being carried from somewhere to the north. Members of the Potawatomi tribe are known to have visited this rock frequently. Some group of early European settlers gave it its current name, thinking it resembled an elephant. To me, it looks more like a stone baleen whale surfacing from the rocky earth.

Kohler-Andrae State Park

Lake Michigan lies in a vast glacially carved basin to the east of Wisconsin. The Niagara Cuesta dips gradually to Wisconsin's eastern shore and into the lake's basin. This makes for a shoreline that is gently sloping, compared to the steep west side of the cuesta that lies inland from the lake.

Kohler-Andrae State Park sits just south of the center point of Wisconsin's Lake Michigan shore. If the wind is blowing a certain way in the park, you can find a sandy area near the lake and watch what the wind does to the sand. Waves bring sand to the shore, where winds dry it and blow it farther inland. If you watch closely, you might see windblown grains of sand landing on the leeward side of a piece of driftwood or a shrub growing near the shore. Watch long enough, and you will see these grains adding up to a small pile that becomes larger with continuing wind. You will be witnessing the birth of a sand dune—the type of landform that is the central feature of this state park.

Like all geological formations, a dune takes time to form, but it occurs in an eye blink compared to the formation of a mountain or a gorge. Grain by grain, windblown sand builds a pile that itself becomes subject to the wind. The pile mounts up asymmetrically, like a water wave, with a sloping windward side and a steeper leeward side, until a tiny avalanche occurs on that steeper side. This moves the pile a few inches inland and in this way, by avalanching, the dune migrates inland, while a new dune begins to form in its place.

As successive dunes form near the lake, they shield the deeper, older inland dunes from the wind. With sands on these older dunes shifting less, certain plants take hold on their surfaces. These pioneer plants, first marram grass and

5.11 The dunes of Kohler-Andrae State Park.

other grasses, further stabilize the dune's surface to allow for more complex ground plants such as creeping juniper, and then shrubs and trees. Eventually, forests and even wetlands develop on old dunes. In unstabilized areas, winds can scour away enough sand to create basins, or bowl-shaped depressions in the sand. Where this wind action exposes the water table, groundwater seeps to the surface creating a place where sedges and grasses can grow. Other plants soon find this moist area and as they grow, die, and decay, soil is formed, and a shallow marsh called a slack develops. Kohler-Andrae State Park is a wonderful place to observe all stages of dune ecosystem development (Figure 5.11).

Another feature of a dune ecosystem, called a sand blow, can occur where the plant cover on a younger dune is disturbed, such as by fire, by animals digging, or by people walking or climbing on the dune. When that happens, the sand is exposed to winds that can then create an ever-widening barren area that can threaten the whole dune ecosystem. The park's extensive network of cordwalks—planks strung together by cords to make boardwalks across the sand—is intended to prevent such disturbance of the dunes.

As for where all that sand came from, that story goes back to glacial times. The numerous ice masses that advanced from the north each plowed up more of the underlying sandstone, pulverizing it and spreading a layer of the resulting sand across the Lake Michigan basin. Outwash streams flowing with each glacial retreat deepened these layers of sand.

When the last glacier was gone, while the center of Lake Michigan held deep deposits of clay and silt, the lake was ringed by the deep mass of sand that waves and wind are now molding into dunes. Because the winds on this part of the continent mostly blow from the west, the dunes on the east side of the lake, on the Michigan shore, are much more massive than those on Wisconsin's shore. Even so, breezes along any of the Great Lakes' shores blow unpredictably from day to day, and when fronts move through and winds become easterly, Wisconsin's dunes continue to shift and grow.

Over the millennia, Lake Michigan's water level has risen and fallen. As the glacier was retreating from Wisconsin, Glacial Lake Algonquin at one time covered all of today's Lakes Superior, Huron, and Michigan. At its deepest, it was about 25 to 30 feet deep over the Kohler-Andre State Park area. About 8,000 years ago, the retreating glacier uncovered an escape route to the northeast for much of that water, and Lake Michigan's level dropped to as much as 300 feet lower than it is today.[18]

As the glacier melted, the compressed land began to rebound, raising the outlet for the ancient glacial lake and causing lake levels to rise again. By about 5,000 years ago, it had risen to the point where the park area again was under about 25 feet of water. This stage of the ancient lake is called Lake Nipissing, or technically, the Nipissing Great Lakes because it involved the upper three of the present-day Great Lakes. Today, evidence of these ancient higher lakeshores exists in the park and elsewhere along the Lake Michigan shore.

Peoples began to inhabit those ancient shores at least 11,000 years ago, according to archaeologists. Human habitation began with the nomadic Paleo-Indians, who hunted the mastodon and other big game along the shores of Lake Algonquin between about 11,500 and 6,500 years ago. By 5,500 years ago, groups of Archaic Indians were fishing the lake's waters using spears and nets, and hunting deer, bear, and smaller game in the dune forests.

About 3,500 years ago, Early Woodland tribes on the Lake Michigan shore added gardening to their survival strategies. They raised corn, beans, gourds, and squash. These people probably were building permanent dwellings and villages by around 3,000 years ago. Later came the Hopewell people, the first builders of burial mounds, between 100 BCE and 1600 CE. Effigy mound builders later joined them.

The earliest Europeans in the area were the French explorers of the 1600s led by Jean Nicolet, Nicholas Perrot, and Jacques Marquette. When the fur trade moved into the area, Indians took part, trading their pelts and furs for clothes, blankets, knives, axes, and other implements. Among the tribes in the area then were the Ojibwe, Menominee, Potawatomi, and Ho-Chunk nations. By 1833, these tribes had ceded all Lake Michigan shore land to the US government and had been forced onto reservations.

From 1850 through 1880, immigrants from Germany and Holland arrived in large numbers, mostly to farm the land inland from the shore. Those who moved to the shore found work as fishers, coopers, and boat builders. At least one farmer tried to raise crops on the land adjacent to the shore within Kohler-Andrae State Park during the 1930s, but it is unlikely that any farmer made a go of working that sandy soil. At one point on the Creeping Juniper Nature Trail, you can see the faint remnants of furrows made by a farmer, as noted on an interpretive sign at that spot.

This park is technically two different parks administered as one. Terry Andrae State Park opened in 1929 on 122 acres of lakeshore land donated by the widow of Milwaukee businessman Frank Theodore Andrae. The second is on 280 acres donated by the Kohler Company in 1966 in honor of the Sheboygan company's founder, John Michael Kohler. Since then the state has purchased an additional 600 acres, to bring the park's total area to more than 1,000 acres.

Those acres comprise sandy beaches along with their fledgling dunes, thick dune forests, and every phase of dune development between those extremes. The park lies in what ecologists call the tension zone—a band of terrain running southeastward across the central part of the state that represents the transition between northern and southern forest types. Many tree species found in the state's northern forests grow in the park because the moist cool conditions created by the big lake are similar to those found farther north. Those trees include white and red pine, yellow and white birch, and certain oak species. Southern forest species are represented in the park by beech, green ash, balsam poplar, cottonwood, and black walnut. The forest also includes the rare plants dune thistle and thickspike wheatgrass, which grow in only a few places on the Lake Michigan shore.

For excellent, detailed, and complete information on all aspects of the park, visit the Sanderling Nature Center, named for the shorebird that scampers near receding waves on the beach poking at the wet sand in search of sand fleas. The center and other park services are supported in part by the Friends of Kohler-Andrae State Park, formed in 1987. The Creeping Juniper Nature Trail Loop departs from, and serves as an outdoor extension of, the nature center. It is well worth the effort to take this easy half-mile hike, which is rich with information on the flora and fauna and history of the park area.

TRAIL GUIDE
The Dunes Cordwalk

Named for the boardwalk strung together by cords that lies flat on the sand, this easy hike parallels the shoreline. Setting out from the northern trailhead at Parking Area 2, the first part of the trail is a flat and easy walk through a dune forest. A little after a half mile, the trail arrives at Sanderling Nature Center, where it merges briefly with Creeping Juniper Nature Trail Loop. From here, hikers take the cordwalk, which continues southwest, ambling over and around dunes in various stages of development—from those just forming to those that are heavily forested, and all stages in between (Figure 5.11). It passes by a variety of dune formations, including large basins, sand blows, and dune forest. These dunes have all formed since Lake Nipissing covered the park more than 5,000 years ago. Over thousands of years, the lake level dropped, exposing the sand from which these dunes were formed.

At about the one-mile point, a spur trail splits off to the right. This short cordwalk takes you to a view of a fledgling dune wetland, or slack (Figure 5.12). Over the coming decades, this young marsh will gain a richer variety of plant life as the area around it becomes dune forest.

TRAIL GUIDE
Woodland Dunes Trail

This easy 1.1-mile loop trail provides an excellent tour of a dune conifer forest ecosystem and includes informative trail signs. It starts from the playground near Parking Area 8. Within the larger loop is a cutoff trail that makes a wheelchair-accessible inner loop. To stay on the outer loop, keep right at the

5.12 A dune wetland, or slack, in an early stage of dune wetland development. Note the creeping juniper bush in the foreground—one of the species that helps to stabilize young dunes.

first trail junction. Heading first inland from the lakeshore area, the trail enters a thick forest sitting over deep sands in soil developed over centuries from the slow, steady process of growth and decay of dune community vegetation. For much of the hike, you walk among towering white pines and red pines. The trail crosses the park's campground, at which point, the Marsh Trail splits off to the right, going to a fully accessible cabin and boardwalk that loops through the Black River Marsh, a well-established dune wetland. The Woodland Dunes trail continues to the left and skirts the marsh for a short distance before turning east toward the lake and returning to the parking area.

Point Beach State Forest

Just north of the center of Wisconsin's Lake Michigan shore is a prominent knob protruding into the lake (Figure 5.2). This is Rawley Point, the site of Point Beach State Forest. Waters off the point are shallow, and the land slopes east from the shore at a low angle. Because of this steady, gentle sloping, a series of parallel ridges and swales have developed on the point as the lake level has dropped over the centuries. It is this feature that gives this state forest its unique character.

In the state forest, the ancient shoreline of Lake Nipissing—25 feet above today's level—is traceable, lying from 0.5 to 1.5 miles inland and roughly parallel to the current shoreline. While the lake sat at that level for centuries, its waves were eroding highlands to the north and south of Rawley Point. Together, water currents and waves carried the eroded sand and built prominent sandbars, or spits, parallel to the Nipissing shoreline.

Eventually, these spits joined to enclose the waters behind them and form a lagoon. Meanwhile the lake level was dropping due to the continuing retreat of the glacier far to the north. As some of the lagoon water drained away, vegetation took it over. Centuries of plant growth and decay formed a deep layer of peat in this wetland, which now lies west of County Highway O on the west side of the state forest. The sand spit that had formed over centuries became the new lakeshore.

The lake level continued dropping, but apparently not steadily. It seems to have stabilized and then dropped, again and again, at irregular intervals. When the lake level was stable for a certain length of time, dunes would build up on the outer beach, creating a parallel low area, or swale, inland and behind them. Over time, grasses and other vegetation would stabilize these dunes and swales, and when the lake dropped again, this process would repeat itself. Older, inland dunes would become forested while their accompanying swales would collect groundwater and precipitation to become wetlands.

What this 5,000- to 8,000-year process created was a series of forested ridges and swales (Figure 5.13), the ridges being between 3 and 26 feet high and the swales filled with peat deposits up to 5 feet deep. The most mature forested dune ridges host juniper, white pine, white cedar, hemlock, red oak, maple, and beech. Some of the swales stay wet, while others dry out by the end of summer. Frogs, salamanders, and mosquitoes all thrive in the wetter areas. State forest literature describes 11 sets of ridges and swales, although some researchers say there are more.[19] Point Beach is one of only three known sites for this formation in Wisconsin. The other two lie farther north on the shore of Door County.

The ridges and swales act somewhat like a diagram of plant succession. The first ridge on the lakeshore is vegetated with plants specialized to grow in beach environments, and the ridges between the beach and the ancient shoreline of Lake Nipissing have undergone progressively longer succession. Thus, the second ridge has been stabilized by junipers, bearberry, and other dune plants, and the dune communities on the next two inland ridges are more diverse. The fifth

5.13 On either side of this swale runs a low ridge that was once a beachfront. This is one of several such ridge-swale formations at Point Beach State Forest.

through eighth ridges host red maple and white birch and a variety of ground plants. The most inland ridges are more fully forested with hemlock, white pine, white cedar, and yellow birch. The eleventh swale, somewhat drier than those closer to shore, hosts black ash and tamarack.

The earliest known human inhabitants of the Point Beach area were part of the Old Copper culture, which refers to early Native American societies that utilized copper for tools and weaponry. Some of these early Native peoples hunted the nearby forests and fished the lake's waters between 5,000 and 3,000 years ago. Other peoples who lived in the area since then included the Potawatomi, Menominee, Odawa, Ojibwe, Miami, Huron, and Ho-Chunk.

In 1835, the US government sold much of the land that is now state forest, previously held by the Menominee tribe, to an Ohio businessman named Peter Rowley. He opened a trading post on the point and traded with the local Odawa and Potawatomi tribes. European immigrant settlers began arriving in the area about 1850. In the later 1800s, the forests in the area fell to the axes of the lumbering boom. Also at that time, the hemlocks were valued for their use in tanning leather. The hemlock groves in and around Point Beach served as a major resource for the booming tannery business centered in nearby Two Rivers.

The shallow waters off the point were bad news for the shipping industry that developed in the eighteenth and nineteenth centuries. Some 26 ships foundered in those waters before the government built a lighthouse on Rawley Point in 1853. There have been no shipwrecks on the point since this beacon was provided. The current lighthouse, which is not available for closeup viewing or touring, was built in 1894. The site of the lighthouse was named Rowley Point, after the former landowner, but since then has come to be known as Rawley Point.

The state forest was established in 1937 on 280 acres of land fronted by Rawley Point. It was originally to be designated as a state park but became a state forest because of legal requirements based on how it was purchased (out of a forestry budget, not a park budget). While the prime motivation for a park was to draw recreationists to the area, the state also conducted extensive forestry on the inland areas away from the point. By 1958, the state and other interests had planted more than half a million red pine, white pine, and spruce trees, with the intention of harvesting them later.

Since 1937, the state has added to the forest through several purchases, and the state forest now includes nearly 3,000 acres with six miles of shoreline. The Works Progress Administration, a Great Depression–era federal jobs program, was employed to build the state forest's Lodge Building, which now houses the nature center and concessions, and the road into the forest. The park also includes three state natural areas (SNAs) set aside for scientific study with limited public access to protect sensitive and rare landforms and species. The Point Beach Ridges SNA is the only known habitat in the state for the sand dune willow.

The state forest also includes 11 miles of hiking and skiing trails, including the Point Beach Segment of the Ice Age National Scenic Trail, and 8 miles of biking trails. A few miles north of the state forest line is the Two Creeks Buried Forest, a fascinating segment of shoreline bluffs that contain remnants of a 12,000-year-old forest. The ancient forest was first flooded and then buried in ice when the retreating glacier re-advanced due to a temporary cooling period that interrupted the larger overall warming period that ended the Ice Age. Well worth a side trip, this area is protected as part of the Ice Age National Scientific Reserve but is accessible for viewing on a limited basis.

TRAIL GUIDE
Ridges Trail, Intermediate Loop

This is a flat, easy hiking and biking trail that traverses a ridge-and-swale formation and intersects with the Point Beach Segment of the Ice Age Trail. It is a loop trail with two cutoffs that create a short (3-mile) loop and an intermediate (5.5-mile) loop within the longest (7.25-mile) loop.

From the Lodge Building parking lot, walk south for about a quarter mile to where the trail crosses a forest road near the campground shower building. From here, most of the hike is on a straight trail along a low ridge with swales on either side. This ridge is located about midway between the beach, where the youngest dunes are forming, and the oldest ridges farthest inland. It is completely forested—thick with hemlock, white pine, cedar, and beech trees. The smell of pine is redolent, and the hemlocks whisper in the wind.

The swales that parallel this trail on either side are also mature. Some sections hold standing water lined with sedges over thickening layers of peat (Figure 5.13). Other stretches of these swales are drier, but for most of the way they are long, low wetlands paralleling the ridges running between them. On parts of the trail, the compacted sand underfoot reminds you that you are walking on an ancient dune that is now being claimed by the forest. Each of the swales on either side of the trail was once part of the lake until a growing sandbar closed it off and became first a dune and later, by the slow process of dune stabilization, the forested ridge you hike on today.

High Cliff State Park

Imagine you are a Ho-Chunk explorer paddling your canoe along the northern shore of Lake Winnebago, the sprawling, shallow lake that now lies in parts of three counties west of the Kettle Moraine. You are traveling toward the northeast corner of the lake and there you find a place to tie up your canoe. You hike inland and begin climbing the high ridge that runs north and south along the lakeshore. After a difficult climb over rocky, ever-steepening terrain, you see through the forest foliage a sight that takes your breath away. A jagged rock wall stands 30 to 40 feet high and runs as far as you can see in either direction along the crest of the ridge (Figure 5.14). At the foot of the uneven wall are broken pieces of it, moss-covered and crumbling into the Earth. Too treacherous to climb, the wall stands as a silent, formidable guardian of whatever lies beyond it.

This wall is the crowning palisade of the Niagara Escarpment. Known locally as the Ledge, it is the central feature of High Cliff State Park. The cuesta's eastern slope at this point is so gradual that it is essentially flat, so the forest beyond the ledge sits on a plateau 225 feet above the level of Lake Winnebago.

The Ledge is made of dolomite deposited by the Silurian sea. Beneath the dolomite are thick layers of shale and sandstone from Ordovician times. Recall that, because of differential erosion of sloping rock layers, the dolomite emerged as a low ridge standing over the eroded sandstone and shale in the basin of Lake Winnebago. Streams formed on the ridge and along its base and continued to wear away at the shale and sandstone lying west of the ridge.

However, were it not for the glaciers, the Ledge would be far less prominent than it is today. The ice sheets flowing from the north followed the lowlands on either side of the Niagara Cuesta and successively plowed away more and more

5.14 The palisade at the top of the Niagara Escarpment in High Cliff State Park.

of the softer bedrock there. Most recently, the Green Bay lobe of the Wisconsin glaciation did much of the excavation of the basin now holding Lake Winnebago. That ice also overtopped the escarpment but did little to the hard dolomite ridge, which now stands higher above the lake basin than it did in preglacial times before the basin was carved deeper.

As the Green Bay lobe melted back, a glacial lake formed within the basin. To the northeast it lay against the retreating ice, and to the east, it lapped against the now higher Niagara Escarpment. This was Glacial Lake Oshkosh, which once occupied all of present-day Lake Winnebago and areas to the northwest, west, and southwest, including Horicon Marsh and the Fox and Wolf River valleys. Between 15,000 and 10,000 years ago, this lake rose and fell at least three times as the glacier retreated and advanced repeatedly at the end of the Ice Age.

As the ice melted over centuries, new outlets for the lake water would open and the lake level could drop quickly and dramatically—as in tens of feet within a few days. When the ice re-advanced, the lake would then rise again slowly as outlets were closed off by newly forming ice. The most recent version of Glacial Lake Oshkosh was about 65 feet higher than today's Lake Winnebago.

When the ice finally retreated and the climate became more hospitable, the earliest human explorers of the escarpment were probably Paleo tradition hunters in search of mammoths, mastodons, and other big game. The earliest evidence of people living in the High Cliff State Park area are its effigy mounds. Archaeologists think nomadic people of the Woodland tradition built these between 1000 and 1500 CE. They are thought to have revered the escarpment, for the mounds lie in the forest atop the ridge just a few dozen yards from the edge of the high cliff.

While some effigy mounds were built in simple shapes such as cones, most were made to resemble animals that lived in the area, including birds, lizards, turtles, panthers, bears, and buffalo. Archaeologists know of 27 mounds that once existed in the park, but only 9 remain. They are protected by law from disturbances of any kind. The balance of the 27 mounds were destroyed by quarrying and other activities within the past few hundred years. The mounds in the park are not known to have contained any human remains. Their exact origins and purposes are unknown, although archaeologists think they may have been used for various spiritual or communication purposes, such as to mark territory, record astronomical events, send messages to a related clan or ancestor, or honor the Earth as part of a ritual.

A series of tribes lived in the High Cliff area, including Dakota, Winnebago (for whom the lake is named and now known as Ho-Chunk), Menominee, Meskwaki (often called Fox), and Sauk. They had all ceded their territories to the US government and were forced off the land by the mid-1830s. By the 1850s, the High Cliff area was attracting European immigrants.

High Cliff State Park not only preserves a piece of the escarpment and its environment, but it also celebrates a colorful past as the site of a bustling mining town founded by the Cook and Brown Company, which made bricks from the clay found in deposits below the cliffs. It began operations in 1855 and built a company store and post office for its employees. This was the kernel of a new village called Clifton, which also served the growing number of immigrant farmers in the area. At some point, the village changed its name to High Cliff because another Wisconsin town called Clifton existed, causing confusion for the postal service. High Cliff grew to be a thriving company town, with company houses, a tavern, and a dance hall. In the 1920s, it even boasted an amusement park operating on top of the bluff.

When the suitable clay deposits were depleted after 60 years, the company shifted to quarrying limestone and eventually sold out to the Western Lime and Cement Company. Dolomite was blasted from the cliffsides, broken into smaller pieces, and hauled down the hill to lime kilns that operated near the lakeshore starting in 1870. These huge ovens, ruins of which are preserved in the park, were used to bake the stone to produce lime, which was sold as an ingredient of cement and plaster. It was also sold to farmers as a soil additive that made their soil less acidic. The dolomite was also quarried to produce building stone and gravel.

The kilns were last fired up in the fall of 1956, by which time the state had purchased the land. By then, the high-quality dolomite had been depleted, as well as the nearby forests, which had provided lumber for the town and firewood for the kilns. The village had dwindled to a ghost town, the company houses and entertainment sites slated for demolition.[20]

High Cliff State Park was opened in 1957. It now comprises nearly 1,200 acres, including more than a mile of shoreline, a 125-acre state natural area designated to protect its unique cliffside ecosystems, and 16 miles of hiking, horse, and biking trails. The Friends of High Cliff State Park was organized in 1997 to preserve, improve, and expand the park by supporting outdoor education and conservation in the park area.

TRAIL GUIDE

Red Bird and Indian Mound Trails

The Red Bird Trail provides a good sampling of the park's environs and views of the lake, and the Indian Mounds Trail takes hikers past the park's nine remaining effigy mounds, for a combined hike of 4.6 miles. The Red Bird Trail, marked with red dots on trees and posts, provides an easy hike on the flat, forested land above the escarpment.

The trailhead is in the parking area near the 40-foot observation tower, which affords a panoramic view of the park, lake, and surrounding area—a good place to start a day at the park. A few yards down the trail, heading south, is the 20-foot statue of Red Bird (Figure 5.15). It is intended to honor the beloved Ho-Chunk chief who, in 1827, was alarmed by reports of the executions of Ho-Chunk warriors near the Mississippi River. In response, Red Bird led raids on white squatters' homes south of La Crosse, killing several family members, and on a barge carrying lead miners and their supplies on the Mississippi near Prairie du Chien. He later surrendered and offered his life to prevent retaliation against his people. Red Bird was highly regarded as an honorable leader by white military officers as well as by the Ho-Chunk. He received a presidential pardon but not before dying in a prison in Prairie du Chien.[21]

About a half mile from the start, the trail enters an abandoned dolomite quarry—a broad area with a flat stone floor and dolomite walls, now being reclaimed by vegetation. At the one-mile mark is an overlook of the lake with benches, beyond which the trail leaves the quarried area behind. Here also is the junction with the Indian Mound Trail, which rejoins the Red Bird Trail farther on. It heads away from the cliffs and into a forest of sugar maple, oaks, and basswood. The trail circles around to head back toward the cliffs and lake, where it passes a series of effigy mounds.

Shortly after rejoining the Red Bird Trail, you will pass an interesting breakaway piece of the bluff. It is a large chunk of dolomite standing separate from the main bluff and completely covered with trees and other vegetation. This sort of formation can be seen all along this trail. The breakaway formations are like towers—stacks of flat rock layers, each of which represents thousands of years of deposition of dolomite (Figure 5.16). Also, look for similar formations that are tipped toward the lake, some of them tipped right over and now crumbling apart. They will soon become part of the lower landscape, as more pieces

5.15 Statue of Red Bird, the widely revered Ho-Chunk chief for whom the trail is named.

5.16 A section of the escarpment has broken away and will eventually fall to the forest floor.

break away above them. In this manner, erosion keeps chipping away at the escarpment, causing it to migrate east.

Shortly beyond the point where the trail comes to the campground, a stairway descends to the right. It is worth taking a side trip down these stairs and among the rocks at the base of the palisade. Step carefully, as the stairs and rocks below can be slippery, especially when wet.

At the 2.2-mile mark, the Red Bird Trail loops away from the cliff and into the woods, curving back to the north-northeast. Shortly beyond that point, it comes to a Y with a bench sitting at the junction of the Red Bird Trail and a well-traveled but unmarked trail to the right. Bear left to stay on the Red Bird Trail as it skirts the campground and heads back toward the trailhead. Much of this part of the trail runs beside historic stone fences—massive, long piles of boulders that border old farm fields. Farmers spent hundreds of backbreaking hours building them with rocks that are hundreds of millions of years old, most of them pieces of the ancient Silurian sea floor.

6.1 Rib Mountain stands 670 feet above the surrounding plain, with a view of Hardwood Hill (left) rising in the distance.

6

On the Bones of the Land

The Northeast Quarter

The major features of the state parks in the northeastern quarter of Wisconsin were formed by a variety of processes. However, what the six parks covered in this chapter have in common is that their foundations are made of some of the hardest rock on the planet. In effect, they lie on the bones of the land.

To know what made northeastern Wisconsin what it is today (Figure 6.2), it is helpful to turn the clock back about 1,500 million years. The Penokean Mountains that had stood for hundreds of millions of years were being worn down to a barren, rocky plain. On the surface, winds and water continued their slow, steady erosive work, sifting sand and gravel from highlands to lowlands, while deep underground more explosive forces were mounting. Magma was rumbling up toward the present-day Wolf River valley area, prying open and filling subterranean caverns, where it would cool to become a massive body of granite. This magma also melted much of the surrounding rock, found its way to several weak points, and erupted to build a row of volcanoes upon deep layers of sandstone and other sedimentary rock. These volcanoes, created about 1,450 million years ago, were centered roughly on what is now Rib Mountain State Park (although Rib Mountain itself is not a volcanic remnant).

During succeeding centuries, as these volcanoes were eroding away, magma that would become a type of igneous rock called syenite flowed up into the rock beneath some of the volcanoes. There it enveloped massive amounts of underlying quartzite—the ultrahard rock that had been metamorphosed from deep layers of sandstone, probably during a mountain-building episode that took

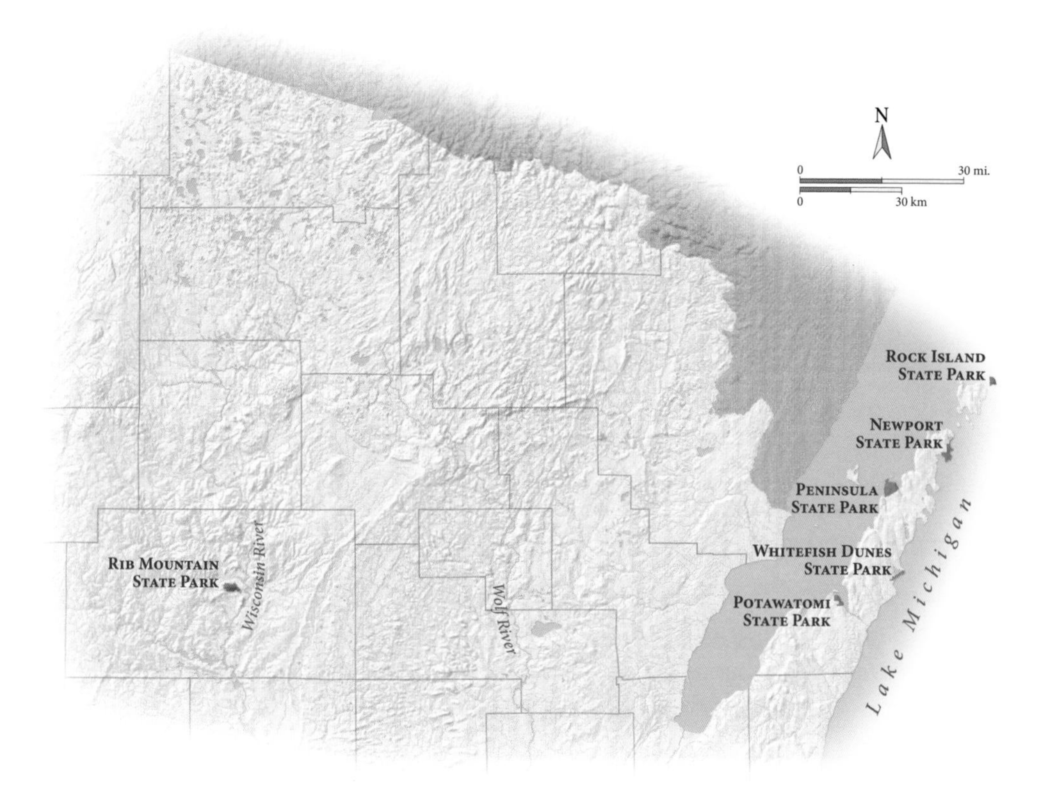

6.2 Northeastern Wisconsin. MAPPING SPECIALISTS, LTD., FITCHBURG, WI

place about 1,700 million years ago. Large blocks of this older quartzite, as well as smaller pieces, were carried up toward the surface by the syenite magma.[1]

The process of volcanic upheaval finally came to an end, and over the coming eons of advancing and retreating seas and intense erosion, these masses of quartzite and syenite would be buried and exposed repeatedly as the softer sedimentary rock was deposited and eroded. They now form prominent hills such as Mosinee Hill, Hardwood Hill (Figure 6.1), and Rib Mountain. These hilltops were eventually buried in ice during the Quaternary period, but not by the most recent glacier. Ancient erratic boulders from earlier glaciers are still found in the area, but most of the older glacial drift has eroded away.

While magma was churning the earth beneath the north-central part of the state, quiet erosion was the prevailing force in the far northeast—in what is now Door County. The Penokeans had eroded to a rolling plain that included all of northern Wisconsin and much of Michigan. By about a billion years ago, north-central Wisconsin was warped up into a gentle dome, slightly higher than the shallow basins on either side of it. The formation of this dome would play an important role in how the land eroded over many millions of years.

By 500 million years ago, a warm, shallow sea covered the area. During the next 60 million years, a series of seas would advance and retreat throughout Cambrian and Ordovician time, depositing layers of sandstone, conglomerate, and shale.

The arrival of the Silurian sea, around 444 million years ago, was a defining event for the area that is now Door County. It was a tropical, salty sea with limited inlets, something like today's Mediterranean Sea, and it covered all of Wisconsin. By this time, ocean life had become far more diverse than before, hosting organ pipe- and horn-shaped corals, antlerlike bryozoans, flowerlike stemmed crinoids, clinging brachiopods, crawling and swimming trilobites, a few primitive fishes, and octopus-like cephalopods that preyed on many of these creatures. Over at least 20 million years, these species flourished, and their remains accumulated on the sea floor and eventually formed limestone that became dolomite. When Silurian seas withdrew, this hard rock covered most of the state in a layer up to 600 feet thick.

Other Paleozoic seas flowed in and retreated after Silurian time, depositing more sediments that became stone. The record of these younger rock layers has been obliterated in Wisconsin by erosion, except for a thin strip of Devonian dolomite lying along the state's far southeastern shore of Lake Michigan. Erosion was the dominant force since Devonian time.

This erosion had an important effect on northeastern Wisconsin. To understand it, think of the state as a generally flat layer cake made of several layers of rock of varying types and thicknesses. Its north-central portion—the Wisconsin Dome—was broadly and gently humped. Wind and rain worked from the top down, wearing away the uppermost layers of rock in the humped area. In this roughly circular area, the lower layers of rock were exposed and likewise eroded, while wind, water, frost, and ice continued to chip away at the edges on the ever-widening circle of eroding upper layers. This effect is still visible today in the rock record and is vividly depicted in Figure 1.1. On that map, note the roughly circular area of tans and browns representing the Northern Highlands, or Wisconsin Dome, surrounded by slightly lower green areas.

Remember that the Wisconsin Dome was gently sloping in all directions away from its center. As erosion wore away the top layers in that central area, the inward-facing roughly circular edges of the eroding layers became steeply sloped. Along these edges, a series of cuestas formed in roughly concentric circles around the dome. Recall that a cuesta is a long ridge with a gentle slope on one side and a steeper slope, or escarpment, on the other (Figure 5.3). Many millions of years ago, these cuestas roughly encircled the dome, but since then, most of them have been eroded away and only remnant stretches are found today.

The most prominent section of cuesta formed by erosion of the sloping land east of the Wisconsin Dome is the Niagara Cuesta, the steep side of which is called the Niagara Escarpment. It remains because it is made of very hard Silurian dolomite, underlain by much softer and more erodible shale. While the softer rock eroded away over hundreds of millions of years, the cuesta resisted erosion, so now stands higher over lowlands on either side of it.

This cuesta underlies all of the Door Peninsula and continues north and east, running around the vast Michigan Basin, which contains Lake Michigan, the lower peninsula of the state of Michigan, and Lake Huron (Figure 6.3). The land surrounding this basin generally slopes inward toward its center, and this sloping, along with the differing degrees of hardness among the rock layers, formed the Niagara Cuesta. On the other side of the basin, this cuesta runs southeast across Ontario to the Ontario–New York border, where the Niagara River thunders over the escarpment at Niagara Falls, for which the cuesta and escarpment are named.

On Wisconsin's Door Peninsula, the Niagara Cuesta's steep west side faces Green Bay. Farther south, on Lake Winnebago, it forms the cliffs at High Cliff State Park. From the top of the escarpment, the Niagara dolomite and underlying rock layers slope east, dipping under the Michigan Basin. This dipping of rock layers means that the western side of the Door County peninsula has high cliffs—part of the escarpment—while the county's eastern shore has more beaches, and its dolomite cliffs are much lower over the water.

The long process of erosion in northeastern Wisconsin was accelerated during the Ice Age. The glacier has been likened countless times to a bulldozer, but another analogy is more appropriate wherever the ice mass ran into high, hard rock such as that of the Niagara Escarpment and Baraboo Hills. In such places, the ice enclosed the rock and continually worked on it. Summer melting allowed water to seep into cracks and crevices where it later refroze, expanded, and fractured the rock. Winter expansion of the glacier caused it to move again, picking up the fractured pieces and carrying them along. In this mode, the

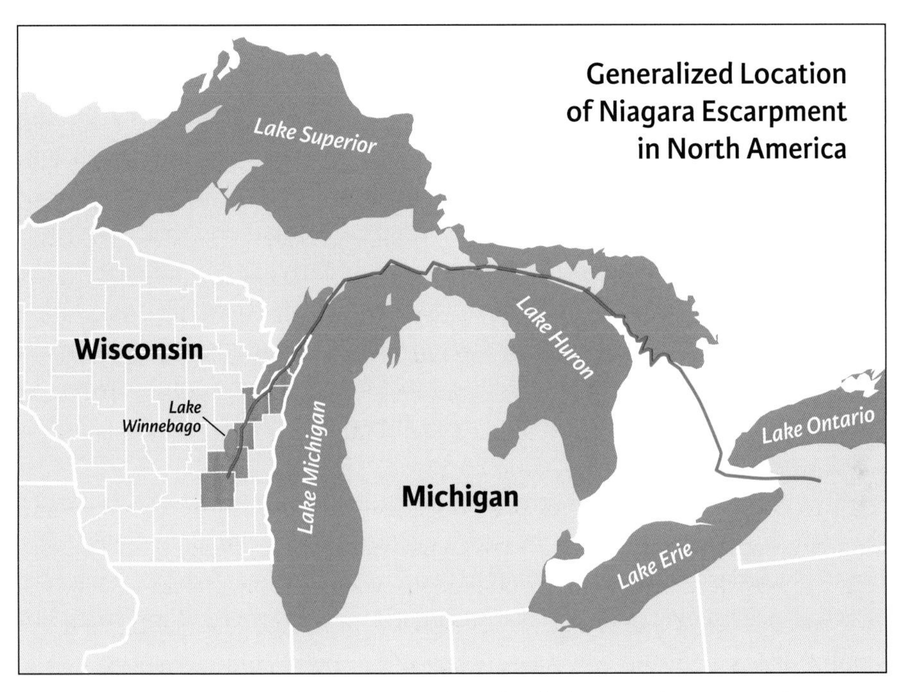

6.3 The Niagara Cuesta rims the Michigan Basin and underlies Door County as well as Niagara Falls. ADAPTED FROM BAY-LAKE REGIONAL PLANNING COMMISSION

glacier was less like a bulldozer and more like an unrelenting wrecking crew, steadily chipping away at masses of hard rock.

The Niagara Escarpment was gradually eroded in this way. It still stands high above Green Bay and Lake Winnebago, partly because the basin to the west was lowered by the glacier. In areas where rock is soft or fractured, glaciers do act as bulldozers. In the basins on either side of the Niagara Cuesta, the glacier easily pulverized and moved the older shale and sandstone, completing the formation of the broad basins that lie under Green Bay, Lake Winnebago, and Lake Michigan.

The ice sheet was so deep and massive that it compressed the land beneath it, and this land has been rebounding since the glacial retreat. Farther north in Door County, where it is thought the ice was thicker, the land was compressed by as much as 300 feet. If it was compressed more than land to the south, it could explain why it seems to be rebounding at a higher rate. Beaches in Door County that formed early in postglacial times are now uplifted 75 feet higher than beaches of the same age 80 miles to the south in Manitowoc County.

The Silurian dolomite that forms the backbone of Door County is still eroding to the east. What remains of the dolomite mass that once covered all or most of the state is a band 230 miles long by 40 miles wide at the most, running the length of Wisconsin's Lake Michigan shore. Part of Door County's geological legacy is due to the fact that dolomite can be dissolved by slightly acidic rainwater. This means the peninsula is laced with small crevices, caves, and sinkholes, all of which began as cracks in the underlying dolomite that were widened and lengthened by such chemical erosion. This type of bedrock, called karst, is especially susceptible to pollution. Water containing pollutants from factories, farms, factory farms, feedlots, parking lots, and septic systems can seep through cracks in the dolomite and contaminate large bodies of groundwater within a few days.

This is just one recent example of how the lives of humans in this part of the state have been shaped partly by its geology over several centuries. The human societies and economies that have occupied northeastern Wisconsin are as varied as its geological story. Native Americans living in the dense forests of north-central Wisconsin had rich resources for hunting and gathering foods, medicines, and hides for clothing. Fur trading was a prominent part of Native economies in the northeast. Some tribes were nomadic, moving from summer hunting grounds to rock shelters or caves for the harsh winters.

Such tribes are thought to have moved around on the Door Peninsula as early as 8,000 years ago. The peninsula's rich resources have drawn people to the area throughout the millennia. These earliest inhabitants found ample forest plants, including berries, nuts, wild tubers, and greens, and hunted deer, elk, beaver, bear, and raccoon. Over the years, Native tribes increasingly fished the peninsula's waters for walleye, trout, whitefish, and lake sturgeon. They found birch bark for making canoes and wiry shrubs to make fishing nets.

The first signs of more settled inhabitants point to a group called the North Bay people, occupying the entire peninsula between 100 BCE and 300 CE. Evidence includes stone tools and pottery. They apparently traveled by canoe and used fishing camps from spring through summer. Their descendants, called the Heins Creek people, occupied the peninsula between 500 and 750 CE. (Between 300 and 500, much of the peninsula was flooded by higher lake levels, so evidence of humans from that time is scarce.) Named for an archaeological site at the mouth of Heins Creek north of Whitefish Bay, they were of the Late Woodland tradition. They continued to develop fishing as a means of survival, in addition to hunting and gathering, and their population is thought to have grown larger than that of any preceding group.

Another Late Woodland group occupied the peninsula between 800 and 900 CE. They are known by their distinctive decorated pottery, more elaborate than that of previous groups, and for their effigy mounds. They extended their seasonal use of fishing villages, staying from spring through late fall, and developed more effective means of catching fish, including gill nets. During winter, they were thought to have sought out the rock shelters and caves of the Door Peninsula. Descendants of this group, the Oneota, appeared on the peninsula around 900 CE. They too improved pottery and fishing methods and introduced agriculture to the area, growing corn and squash. It is thought that with a greater variety of foods and better food storage, they may have been the first people to stay year-round in their camps, thus beginning the establishment of permanent villages.

The Oneota are believed to be the ancestors of the Ho-Chunk people, who have had a major influence on the peninsula's history. The Potawatomi were the dominant tribe on much of the peninsula when Europeans first immigrated, starting with traders who brought knives, firearms, kettles, cloth, and other coveted items to exchange for furs and pelts. The name Potawatomi translates to "Keepers of the Fire." Other tribes that have played major roles are the Ojibwe, Menominee, Sauk, and Odawa tribes.

Some of the earliest whites in the area were French fur traders. A trade route was established from the Straits of Mackinac on the north end of Lake Michigan to the mouth of the Fox River on the south end of Green Bay. The route hugged the north shore of the lake along what is now Michigan's Upper Peninsula, then crossed into Green Bay and followed the western shore of the Door Peninsula the full length of the bay. It was called the Grand Traverse—part of a major connection between European markets and the middle of North America for more than 200 years, through the end of the nineteenth century.

European immigrants made intensive use of the peninsula's resources, notably during the lumbering boom of the late nineteenth and early twentieth centuries when the huge pines of the northern forests were cleared. Door County's climate was well suited to cherry and apple orchards that quickly expanded and flourished. Today, they are still an important part of the county's economy. In north-central Wisconsin, because of the massive presence of granite in several deposits, granite quarrying became a thriving business starting in the late

1800s. With the coming of Europeans, the Native tribes were steadily edged off the land. In the 1830s the Menominee, Potawatomi, and other tribes ceded their lands in northeastern Wisconsin.

Since the 1800s, other industries that have thrived on the Door Peninsula are commercial fishing, shipping, and ship building. Tourism also started in the later 1800s and is now an important part of the economy throughout northeastern Wisconsin. Travelers and day-trippers who seek a deep forest experience in any season can find it in the northeastern quarter west of Green Bay. Those who prefer sailing, exploring pristine islands, hiking on lakeside cliffs and shores, or touring quaint old lakeside towns are drawn to Door County. As in the other parts of Wisconsin, all of these experiences can be sampled in and around the region's popular state parks.

Rib Mountain State Park

Imagine that you could turn the clock back about 1,500 million years and travel to the center of what is now Wisconsin, heading west. Approaching the current site of the city of Wausau, you would see before you a row of four volcanoes stretching from north to south.[2] You might see them belching smoke and ash and sense the Earth rumbling underfoot, for beneath those peaks, magma would be moving, heaving quartzite forged from sandstone upward toward the base of the volcanoes from deep under their foundations.

If you could hang around for about a billion years, you would see those volcanoes become dormant and gradually shrink away. During that time, syenite magma was heaving up blocks of 1,700-million-year-old quartzite beneath the volcanoes. Wind and rain would whittle the volcanoes down from heights of many thousands of feet until they would essentially disappear into the rolling rocky plain spreading across the surrounding area. However, a set of freestanding hills would remain in their place—tiny with respect to the original total mass of the volcanoes, but still imposing. Those hills, called monadnocks, were made of nearly pure quartzite lying within a ring-shaped area inside the footprint of one of the vanished volcanoes (Figure 6.4).

The erosive forces that dismantled the volcanic peaks were no match for the ultrahard quartzite that stood firm against eons of erosion. Over the next 300 million years, the quartzite peaks would be submerged under shallow seas, several times, and buried under layers of sedimentary rock that would be carried away just as the volcanoes were by flowing streams and winds. Probably more than once, the stalwart quartzite remnants, while being reduced from their original sizes, withstood the erosion, while softer sedimentary rock surrounding them was carried away.

When glaciers first entered Wisconsin around 2 million years ago, one or more of the four major ice sheets engulfed the monadnocks. However, the most recent one stopped short of these quartzite hills. To the east, the wall of the Green Bay lobe stood about 15 miles away for about 3,000 years, while the Wisconsin Valley lobe advancing from the north stopped less than 25 miles away. In the area surrounding the monadnocks, geologists see no drift that can be sourced to the most recent glacier, but very thin samples of much older drift, primarily boulders, are scattered across the region. Pieces of quartzite from the monadnocks have been found west and south of their location. All of this tells geologists that earlier glaciers did in fact cover the area.[3]

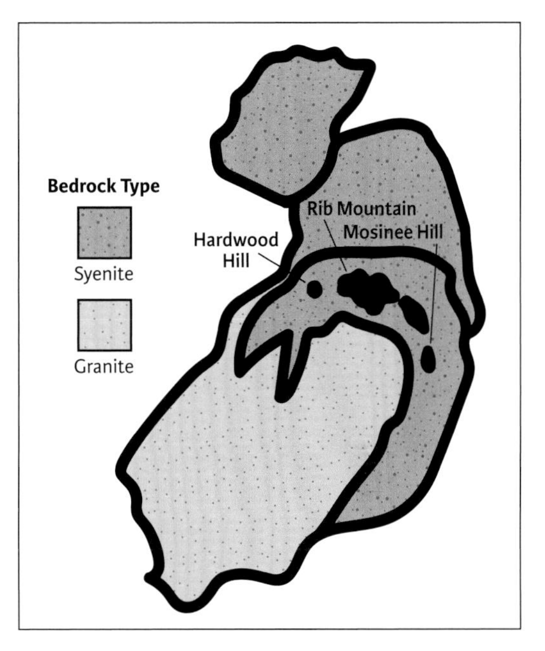

6.4 Footprints of volcanoes that once stood in the Rib Mountain area. The southernmost area lies over granite bedrock, and the other three over syenite, which uplifted the quartzite peaks called Rib Mountain, Hardwood Hill, and Mosinee Hill, within the southernmost syenite body. ADAPTED FROM LABERGE, *GEOLOGY OF THE LAKE SUPERIOR REGION*

What the glaciers left behind was a series of quartzite hills, the most prominent of which are three high hills that lie in the roughly ring-shaped area—Hardwood Hill, Rib Mountain, and Mosinee Hill (Figure 6.4). The largest of these three is Rib Mountain, a slightly curving ridge 4 miles long and up to a mile wide lying roughly east to west. It was named for its rib shape. The Ojibwe too saw this resemblance, calling the hill Opicwana, meaning "rib." For a long time, it was called Rib Hill, but in 1949 its official name was changed to Rib Mountain, possibly with the intention of drawing more tourists.

Rib Mountain's core is a block of quartzite at least two miles long and a half mile wide, reaching 760 feet above the Wisconsin River, which flows past the hill's east end. Geologists believe this quartzite block, most of it still embedded in the syenite below, is much larger, perhaps several thousand feet thick.

While Rib Mountain is the fourth-highest peak in Wisconsin (just 28 feet shorter than the highest, Timm's Hill), it is the highest exposed bedrock in the state, which gives us a sense of the power of the magma that lifted it to its present height and far higher. Its long, slender summit of quartzite is steep-sided for most of its length, with much of its base draped in talus, the broken chunks of quartzite fractured from the main mass by frost and dropped in heaps. The hill below the summit and talus is more gently sloping but still steep, and is composed of syenite and other more erodible forms of rock.

The mountain of pinkish to purple quartzite was once seen as a massive resource. Starting in the 1870s, a sizeable chunk of the mountain's west end was quarried and hauled away, mostly to provide dimension stone for building and landscaping materials, monuments, and gravestones, and to be crushed and used for making abrasives. The state park was established in 1929 but did not include the land around the quarry. The quarry was closed by 1990, and the state obtained the land and added it to the state park in 2000. The nonprofit Friends of Rib Mountain State Park was formed in 1993 to assist the Department of Natural Resources in preserving, improving, and promoting the state park. The Friends group was responsible for developing and mapping the park's hiking trails, building its 200-seat outdoor amphitheater, acquiring additional land, and many other projects.

The state park is unique, as is Rib Mountain itself, for sitting higher above the surrounding land than any other site in Wisconsin. From its 60-foot observation tower and from various points along its highest trails, you can view the neighboring quartzite peaks Mosinee Hill (actually two hills joined by a saddle) to the south-southeast and Hardwood Hill to the west (Figure 6.1). To the northeast, you see Wausau and farther east the Johnstown Moraine—the low north-south ridge that forms the western extent of the Green Bay lobe. Between the park and moraine are outwash plains where water flowed away from the melting glacier for thousands of years. On the clearest days, you can look north and see the moraine deposited by the Wisconsin Valley lobe. Between the park and that moraine are low hills formed by much older glaciers and now dissected by the Wisconsin River and its tributaries.

Part of the southwest third of the park was designated as the Rib Mountain Talus Forest State Natural Area, which protects micro-habitats on the talus slopes where rare plant species, including purple clematis (*Clematis occidentalis*), Missouri rock-cress (*Arabis missouriensis*), and the state-threatened drooping sedge (*Carex prasina*), have been found. A few of the park's trails cross this area, and here especially, hikers are asked to stay on designated trails so as not to disturb rare habitats. The park has a total of 13 miles of trails, 8 of those miles accessible to people with disabilities.

Three of the trails, the Red, Green, and Blue Trails, were built by the Civilian Conservation Corps (CCC) in the 1930s. CCC workers cut and placed the extremely hard and heavy quartzite stones into the several stairways on the trails using no power tools or motorized vehicles. They also built a picnic gazebo, the original downhill ski runs on the northeast side of the hill, and the ski chalet that is used today.

TRAIL GUIDE

Red Trail

Of the several trails on Rib Mountain, this one provides the best sampling of the park's features, skirting the base of its rocky summit with good views of the surrounding landscape and of the quartzite cliffs and talus fields. Except for the rocky stretches that are slippery when wet, this is a moderately easy trail, just under two miles long, with a few steep sections. Several stretches are asphalt-covered and incorporate stone steps. The trail is well marked but often changes direction suddenly, so keep looking for the red trail markers.

The trail heads southwest from the concession stand near the observation tower on top of Rib Mountain. It begins with a descent off the top of the mountain, meandering gently downhill and clockwise around the summit. After a half mile, you pass the low point on the trail and begin a gradual ascent through a thick birch forest. Shortly beyond this point, the trail climbs across a talus slope—a massive collection of quartzite boulders that have accumulated over thousands of years, most of them pried loose from the quartzite mass above by frost. At 0.75 mile, the trail has reached a high point with a wide-open view to the west and Hardwood Hill (Figure 6.1). While viewing the rocks and broad plain below, you might take a moment to contemplate the incredibly long time it took for this mound of quartzite to form and to be heaved up as a bulwark against centuries of erosion.

At the 1.3-mile mark, a striking rock formation sits to the right—a chunk of layered quartzite that was heaved up and turned on its side (Figure 6.5). Outcroppings such as this provide evidence of dramatic upheaval of the rock

6.5 This boulder, made of layers of quartzite, was heaved on its side during the uplifting of Rib Mountain nearly 1.5 billion years ago.

layers probably more than once in the distant past. They were twisted and tilted violently by the force of moving magma. Once buried deep beneath other layers, pieces of these rock masses now sit atop Rib Mountain.

From here, the trail curves to the north. At 1.5 miles, you pass a moss-covered talus field on your right, remarkable because it is probably much older than other talus slopes in the park. Shortly beyond, the trail reaches its northernmost stretch with panoramic views of Wausau and the surrounding area. In another tenth of a mile, the trail begins skirting the edge of the ski hill area and climbing back toward the top, where a resting bench affords a view of the Rib River, which flows into the Wisconsin River just east of Rib Mountain. Once the trail reaches its junction with the Green, Gray, and Blue Trails, it crosses the park road and heads back to the concession stand.

Potawatomi State Park

About halfway up the Door Peninsula on the Green Bay side, the waters of Sturgeon Bay nearly bisect the peninsula (Figures 6.2 and 6.6). The deep valley holding the bay penetrates eight of the nearly ten miles from Green Bay to the Lake Michigan shore. (A canal now connects the bay with the big lake.) The story of Sturgeon Bay begins with the retreat of the last inland sea from Wisconsin and the exposure and drying of the dolomite mantle that was left covering most of the state. Between 400 million and 300 million years ago, the sunbaked dolomite cover developed cracks, along which streams eventually began to flow. Over time, slightly acidic water in precipitation dissolved and carried away some of the stone, widening and deepening the new valleys.

Before the glaciers came, the Niagara Cuesta that now defines Door County was less prominent than it is now, as were the basins on either side of the cuesta and the short stream valleys flowing across it. With the coming of the

6.6 A view from the former observation tower at Potawatomi State Park. Sawyer Harbor is bounded by Sherwood Point, beyond which lies Sturgeon Bay.

Quaternary period, each advance of the glaciers inching across the land dug the softer stone in the basins deeper and carved away some of the harder stone on the peninsula, especially where erosion had started in the valleys such as that of Sturgeon Bay.

During each glacial retreat, meltwater continued this erosive process. Geologists think that the Sturgeon Bay valley might have been the location of a subglacial stream at some point, which would have done more carving even as the ice still sat on the peninsula.[4] It is thought that, for some amount of time late in the last melting period, Glacial Lake Oshkosh (p. 185) drained through what is now Sturgeon Bay, boring the valley wider and deeper over decades.

By the end of the Wisconsin glaciation, the Door Peninsula stood starkly above the flanking basins. On both sides of the mouth of Sturgeon Bay rose polished white dolomite headlands—each part of the now relatively higher escarpment that stretched away to the north and south along the Green Bay shore. Many centuries later, the headland standing over the south side of the bay would become the site of Potawatomi State Park.

Since the glacier retreated, lake levels on the peninsula have changed a number of times. Beaches formed on Glacial Lake Algonquin around 11,000 years ago, and on Lake Nipissing about 5,500 years ago. Remnants of these ancient beaches are visible in Potawatomi State Park today, well above the current lake level. Whenever the lake sat at a higher level for decades or centuries, its waves and currents would wear away at the lakeside cliffs, eroding them inland and creating terraces along the shore. Today, park visitors pitch tents and park their RVs in the campgrounds located on one of these terraces.

As part of the ancient shore erosion, the waves of Lake Algonquin cut caves into the dolomite cliffs on the north and east bluffs in the park. Because of the drop in lake levels and the rebounding of the glacially compressed land, these former seaside caves now sit as high as 60 feet above the lake. A few can be seen from the road to the boat landing on the north side of the park, but they are not easily or safely accessible, and park officials keep them off-limits to exploration.

Evidence indicates that the earliest humans in the park area were Late Woodland people who used the lakeshore just west of the park as winter fishing grounds.[5] The Potawatomi, for whom the park is named, and their forebears caught sturgeon and northern pike and stored them by packing them in cold, wet sand in pits within their camps.

In the 1800s, dolomite was a popular building material, so Sturgeon Bay with its harbors and massive exposures of dolomite became a prime site for the quarrying and shipping of the building stone. By 1900, the stone business was booming in the bay. Before the boom, the US government had undertaken some quarrying within what are now park boundaries, near Quarry Point. In 1837, the US War Department acquired the land to locate a fort, viewing it as a strategic site for the remote possibility of a war with Canada.[6] The war never came and the fort was never built, but the quarry did yield stone that the government used to build breakwaters for the major harbors on Lake Michigan.

What became known as Government Bluff later drew the attention of conservationists. Some were upset with the fact that logging at the turn of the century—much of it done by squatters with no legal claim to the timber—had depleted the area's forests. The conservationists and others succeeded in getting the state to buy the land and establish a park in 1928. It was named Nicolet State Park, for the first European explorer to visit the area. The name did not last long, however, as the head of the state's Conservation Department asked the people of Sturgeon Bay to suggest other names. Potawatomi was chosen in honor of the tribe that had lived in the area before ceding its land to the US government in the 1830s.

The park now comprises about 1,200 acres and nearly 3 miles of shoreline on Sturgeon Bay, with 8.5 miles of hiking trails, 8.5 miles of cross-country skiing trails, and 8 miles of biking paths. It hosts forests of oak, maple, pine, basswood, and beech, while ancient white cedars cling precariously to the cliffs. You can also view protected plant species, including trilliums and lady's slippers. The woods are rich with animal life: deer, raccoons, foxes, opossums, and at least 50 species of songbirds. Keep your eye out for the pileated woodpecker, with its flashing black-and-white feathers and brilliant red head—a stirring sight if you're lucky enough to spot it.

TRAIL GUIDE
Tower Trail

This is an easy trail, mostly flat and wide and crossing through beautiful forestland, beginning from the Old Ski Hill Overlook in the northwest corner of the park. The overlook provides a panoramic view of Green Bay to the west and Sawyer Harbor to the north. From there, the trail runs near the edge of the plateau atop the bluff, but it is heavily forested with no open views.

Until 2018 visitors could climb to the top of a 75-foot tower, sitting atop a 150-foot bluff, for spectacular views of the forest behind you and of the bay and harbor to the north (Figure 6.6). Over centuries, the waves and currents flowing around the mouth of Sturgeon Bay moved sand to create the hook-shaped peninsula that forms Sawyer Harbor. The major part of it, oriented east-west, was formed around 5,000 years ago when Lake Nipissing was at its high point. It is called Sherwood Point. When the lake dropped, currents changed and formed the smaller sand spit, called Cabot Point, angling southeast off the east end of Sherwood Point. From the old tower, built in 1932, you could also see the abandoned Leathem and Smith Quarry straight across Sturgeon Bay. This was an immense quarry, from which stone was moved to most of the major ports around Lake Michigan to protect the harbors.

At the site is the eastern terminus of the Ice Age National Scenic Trail, which runs with Tower Trail for almost two miles. Along the trail, slabs of rock crop up in places, revealing how thin the topsoil is over the dolomite bluff. At about 0.8 mile, the trail begins to descend toward the bay, crossing over low

6.7 The Tower Trail in Potawatomi State Park runs along a lakeshore terrace carved by Lake Nipissing.

ridges that are remnants of ancient beaches where the lake once lapped the shore and rolled the rocks there to round them off. They are now covered with moss as the forest overtakes the old beach.

Shortly beyond the one-mile mark, the trail crosses Shoreline Road and descends steps to the shoreline terrace that was created by waves of a higher lake centuries ago. You will see here the layered nature of the dolomite that was deposited by succeeding advances of inland seas hundreds of millions of years ago (Figure 6.7). Now water, frost, and vegetation are steadily dismantling the layers. The park's visitor guide aptly describes the hike on the flat shoreline terrace as a walk "through an archway of cedars."[7] These trees appear to be growing right out of rock in many cases. They have an ability to extend their probing roots into the smallest nooks and crannies among the rocks.

Near the two-mile mark, Tower Trail splits off to the right, while the Ice Age Trail proceeds ahead along the shore. Tower Trail then crosses Shoreline Road again and enters the north campground. At 2.5 miles, it turns west and climbs off the terraces, once again crossing an old beach ridge—this one built by the icy waves and currents of Lake Algonquin as the glacier was retreating to the north. After the trail turns from westerly to northerly, it meanders for another mile to close the 4.3-mile loop at the Old Ski Hill parking area. A cutoff trail is available on the shoreline terrace to make the hike a mile or so shorter.

Whitefish Dunes State Park

Across the peninsula from the high cliffs overlooking Green Bay, Door County's east shore sits lower over the waters of Lake Michigan. Here the cliffs are less spectacular, but this side of the peninsula has a pattern not found on the Green Bay side. Look at any map of the peninsula and you will notice, starting at the northern tip and running all the way to Sturgeon Bay, a series of inlets, some of which have been closed off by narrow strips of land to form small lakes close to the Lake Michigan shore. Such is the case with Whitefish Dunes State Park (Figure 6.8).

The story of the park as it appears today starts with the Silurian sea, which deposited the thick layer of dolomite that makes up the 10- to 20-foot cliffs at the north end of the park. When the sea departed, the sea floor dried and cracks formed in the dolomite mantle. Because dolomite is dissolved by water,

6.8 Low dolomite cliffs lie over Lake Michigan in the northern part of Whitefish Dunes State Park.

streams flowing along these cracks widened and deepened them, gradually washing away some of the dissolved stone. Because the land is tilted slightly to the east, the longer streams on the peninsula flow south-southeast. Water melting from the glaciers would have roared down these valleys, widening and deepening them further. Where they entered Lake Michigan, these streams carved the inlets that you now see on the map (Figure 6.2). The southernmost of the eastern inlets, now called Clark Lake, was to become the site of Whitefish Dunes State Park.

Geologists think that about 5,000 years ago, what is now Clark Lake was part of a bay with its mouth lying between Cave Point on the north end of the park and Whitefish Bay Creek on the south end. Waves and currents in what was then Lake Nipissing swirled off Cave Point, moving sand and building a spit that stretched southwest from the point. This sandbar grew and was eventually exposed when the lake level dropped. Currents kept adding sand to the sandbar and as more of it was exposed, dunes began to form.

Dune formation takes decades to centuries (see also Chapter 5, p. 174). The wind blows sand grains inland and where they encounter an obstacle such as driftwood, they drop on the lee side of the object and start mounting up. A dune is built in this way, grain by grain, gradually migrating away from the shore. At some point a new dune begins to form in front of it, as pioneer plants such as marram grass take hold on the older dunes, which then become stabilized by grasses and other forms of vegetation. Slowly dropping lake levels make room for more new dunes as the older dunes continue migrating inland. With centuries of plant growth and decay, soil gradually forms on the older dunes and they can support increasingly complex plant communities and eventually dune forests.

In this way, the strip of land that makes up most of this state park, lying between Clark Lake and Lake Michigan, was formed. While dolomite bedrock is close to the surface in most of Door County, you could dig for a long time in the center of the park before reaching it. All that separates Lakes Clark and Michigan is deep, heavily compacted sand and the forest that grows on it.

Much of what we know about Native American history on the Door Peninsula comes from archaeological work done at Whitefish Dunes State Park. The park hosts multiple ecosystems, including rocky shorelines; all stages of dune formation, from young and fragile to old and stable; dune conifer forests; meadows; wetlands; and inland hardwood forests. They in turn provide a variety of habitats for plants and wildlife, which drew people of the Middle Woodland tradition to the area at least 2,100 years ago.

Archaeologists working in the park have found evidence of several distinct groups occupying a village site in the park since then. The first such identified group were the North Bay people, beginning around 100 BCE, followed by their descendants, the Heins Creek people. These tribes found fishing to be especially good in the park area and the village began to grow. By the time the people of the Late Woodland culture occupied the site between 800 and 900 CE, this village was well established. The Oneota occupied the park area beginning around 900 CE and developed agriculture, which allowed them to stay year-round in the village. Near the park's visitor's center are reconstructed lodges and interpretive signs that inform visitors about Native American life in the area.

Like many of their predecessors, European immigrants took to fishing in Whitefish Bay. In the 1830s, when commercial fishing was beginning, the bay was called Fisherman's Bay. It served as a small port through which fishing supplies, barrels of salted fish, and lumber were shipped. The bay was the site of six shipwrecks, and the remains of one wreck are displayed in the park just as they were found on the bottom of the lake by a salvage crew. The Whitefish Bay port was abandoned after the lumbering boom ended in the early 1900s.

By the 1930s, scientists were beginning to realize the ecological value of the Whitefish Dunes, Wisconsin's largest and most developed sand dunes. Even the most stable dunes are fragile ecosystems, and visitors to the sandy beach were taking a toll. In 1967, the state park was established to protect 867 acres of dunes and dune forests. A little over a quarter of the park land has been designated a state natural area to be left completely undisturbed, in order to protect the rarest of dune flora and fauna, including the state's largest population of the federally threatened dune thistle.

Within the park's boundaries on its northern end is Cave Point County Park, an area of a little more than 40 acres. The lakeshore side is a spectacular collection of water-worn cliffs and caves—well worth the short drive or hike from the Whitefish Dunes Nature Center. Also departing from the Nature Center are a series of trails totaling 14.5 miles in length, one of which passes three access points to the park's 1.5-mile sandy beach. The Nature Center itself contains a wealth of information about the geological and natural history and the daily lives of Native Americans who lived in the area between 2,100 and 200 years ago.

Visitors find an outdoor extension of the Nature Center on the Brachiopod Trail, named for a particularly compelling part of the park's geologic and natural history—an ancient creature that you might say lives on in the rock record. Clamlike in appearance but not at all related to present-day mollusks, the brachiopods dwelled in the shallows of the Ordovician and Silurian seas. They were so abundant that their remains are a major component of the dolomite that now makes the backbone of the Door Peninsula.

This 1.5-mile flat loop trail departs from the Nature Center where you can pick up a detailed, informative pamphlet that serves as a trail guide keyed to numbered stops on the trail. It explains the ancient history and ecology of the area and helps you interpret what you will see on your hike. The trail first takes you to views of the low dolomite cliffs on the bay and the arc of sandy shoreline in front of the dunes (Figure 6.8). It then meanders through the upland hardwood forest and over a wetland that connects to Clark Lake, which is accessible via a spur trail, before circling back to the Nature Center.

TRAIL GUIDE

Red Trail

This three-mile moderately easy loop trail provides the best sampling of the shoreline, dune, and dune forest environments. It starts from the Nature Center and after about 0.2 mile, reaches the first of two beach access spur trails (marked Beach Access 2) on your left. This is a boardwalk with steps up and over the large front dune, which is designated as "active" because it is in the early stages of stabilization. Hikers are asked to stay on the boards to help preserve this fragile system, but you can keep going to the beach where you can roam on the shore.

The Red Trail continues southwest, winding among heavily forested old dunes, some of which, as they rise beside the trail, seem dizzyingly steep, with large trees on top making them seem even higher.

At 0.8 mile, the second spur trail to the beach (Beach Access 3) splits off to the left. Here also, the Green Trail splits to the right and meanders through dune forest back toward the Nature Center with two shorter trail options. Continuing, the Red Trail passes through thick stands of white cedar and balsam fir with white pines interspersed.

At 1.2 miles, the trail does a right turn to a set of wooden steps. (Straight ahead is a short spur that connects with a town road at the park boundary, not part of the Red Trail.) The steps go up and over another established dune and traverse it on a well-built boardwalk and cordwalk. Shortly beyond the 1.3-mile mark, another set of steps and boardwalk go right toward Old Baldy—the highest dune in the park and one of the highest in the state. It is another 200 yards to the top of Old Baldy, which sits 93 feet above the lake level (Figure 6.9).

The top of this established, yet fragile high dune is protected by a platform with views of Lake Michigan and Clark Lake. As the crow flies, the big lake is less than 300 yards to the southeast of the platform. The smaller lake, the ancient inner bay, is about a half mile to the northwest. This viewing platform is an excellent place to sit and enjoy the lake breezes while pondering the incredibly slow process of dune formation and the centuries it took for this amazing set of dunes to develop and become covered by a diverse forest. Your perch on the platform is situated 75 feet above what 5,000 years ago would have been the icy waters at the head of the ancient bay, which stretched for miles to your northwest well beyond today's Clark Lake shoreline.

6.9 Nearing the top of Old Baldy, the highest sand dune in Whitefish Dunes State Park.

Back down on the Red Trail, you curve around the back side of the big dune and head northeast. This route back toward the Nature Center is less hilly and makes for a delightful and easy inland hike among pines, junipers, birches, and beeches. If you are quiet and watchful, you will spy plenty of birds and maybe a deer bounding away, all the while enjoying the sound of the waves of the big lake washing the shore to the east.

Peninsula State Park

The Niagara Escarpment lies on the Green Bay side of the Door Peninsula and perhaps nowhere is it on display more spectacularly than at Peninsula State Park. Gleaming white cliffs of dolomite tower as high as 150 feet over the water and stand starkly above inland forested areas (Figure 6.10). The escarpment generally arcs north from Fish Creek, east through the northern part of the peninsula, and then south toward Ephraim, bounded on either end by stream valleys.

6.10 These dolomite bluffs stand more than 100 feet above the waters of Green Bay in Peninsula State Park.

Those stream valleys began as cracks in the dolomite mantle that lay over the land 400 million years ago. The cracks were etched by rainwater that slowly dissolved some of the alkaline rock, gradually deepening them. The glaciers scoured them wider and deeper, and meltwater streams gushing into the lake scooped out the Fish Creek and Ephraim bays, leaving the dolomite peninsula jutting into Green Bay. By the time the ice melted back 10,000 years ago, the setting looked much as it does now, minus the forest cover.

On the postglacial shores of the state park's peninsula, waves and currents worked on the cliffs for centuries, pulling pieces of the cliffs down and smashing and rolling the fragments, rounding them off to make the dolomite cobblestones that now lie along the shores. On the inland cliffs, winter frost, summer rains, and creeping vegetation picked at the rock, gradually breaking off sections of the cliffs to create talus fields at their bases. In this way, the spectacular palisades of the park were formed and continue now to be sculpted.

On a map, the knobby, small peninsula extending from the north side of the main peninsula looks a little like a dog's head, collared by Tennison Bay to the east and Nicolet Bay to the west. Behind the dog's ear sits the Eagle Bluff Lighthouse, and the dog's snout is pointing toward Horseshoe Island. This little peninsula is an outlying dolomite mass. It was an island in Glacial Lake Algonquin in early postglacial days and possibly also in Lake Nipissing about 5,000 years ago.

The waves of the glacial lakes battered the cliffs and exposed segments of the flat layers of dolomite, creating shoreline terraces around much of the peninsula. In some parts of the park, these terraces resemble giant steps leading from the shore to the bluffs. For example, you can see this formation by looking north from Sven's Bluff to the headland across the water. On the terrace carved by Lake Nipissing sit the town of Fish Creek, the Tennison Bay playground, the Nicolet Bay store, and the Village of Ephraim.

Another sign of heavy wave action along the shores of the glacial lakes is the existence of caves in the cliffs. Once such an opening got started, persistent icy waves kept working on the joints and cracks in the layered dolomite. Because of the way this rock fractures, the caves were sharp-angled openings, gradually made deeper by the rhythmic wave action. When water levels dropped and the land rebounded after being compressed by the glacier, the caves slowly rose away from the shorelines. A stark example of this is Eagle Cave, a large rectangular opening 30 feet above the water, carved into the park's northeasternmost cliff (Figure 6.11).

The spectacular rock formations, cave shelters, forests, wetlands, and lake waters rich in plant and animal life attracted people of the Early Woodland culture, the earliest known visitors to the park's peninsula, at least 2,500 years ago. First came nomadic tribes that moved from season to season. Later groups eventually established more permanent villages. Archaeologists found evidence of a village site dated to 600 to 400 BCE at Nicolet Bay Beach where today's visitors go for sunning and swimming. Several Native tribes have occupied the site since that time. In later years, the most prominent tribe was the Potawatomi.

Early European immigrants also hunted and fished and traded with local tribes. The Grand Traverse trading route on Lake Superior brought travelers past the park's peninsula, and it is likely that they used Eagle Harbor just east of it as a place to rest and restore food supplies. Later came the loggers who cleared most of the park's forests and several families who moved in to farm the limited areas of the peninsula that were suitable for crops and grazing, such as the flatter

6.11 Eagle Cave is a well-known example of a water-cut cave created thousands of years ago when the lake level was higher.

areas near Weborg Point and Nicolet Bay. They grew oats, barley, wheat, rye, and corn and established cherry and apple orchards.

In 1868, the federal government erected the Eagle Bluff Lighthouse as part of an ambitious program to light the shores of the entire Door Peninsula, where shipwrecks were becoming more frequent as industry and trade grew in the area. By the 1890s, the peninsula had become a tourist destination, and conservationists, led by farmer and state assemblyman Tom Reynolds, began to argue for protecting the natural features of the area from uncontrolled development and exploitation. At the same time, those hoping to benefit from tourism also saw value in creating a park on the peninsula. In response, in 1910, the Wisconsin legislature established a state park there—only the second after Interstate Park.

The Civilian Conservation Corps (CCC) came to the park and built Camp Peninsular in the 1930s. Over the next several years, CCC workers built stairs into some of the steeper trails, cleared new trails, improved roads, and planted trees. They also built a 1,200-foot toboggan run that is now being reclaimed by the forest but is still visible near the park's Nature Center. The CCC's tenure in the park was controversial, partly because the park superintendent who had done much to develop the park seemed to regard the arrival of the CCC as an invasion of sorts. Others with more of a focus on preservation objected to the CCC work, referring to is as an attempt at "manicuring the wilderness." Nevertheless, the CCC made certain trails and other features more accessible and safer for the growing stream of tourists, and generally they were highly regarded by residents and visitors to the park.[8]

Today, Peninsula State Park comprises 3,776 acres—the third-largest park in the state—with eight miles of shoreline and two specially protected areas. The White Cedar Forest and Beech Maple Forest State Natural Areas (SNAs) each hold native plant communities that closely resemble those of presettlement times. The park also encompasses Horseshoe Island, the picturesque isle visible from several overlooks in the park. It has a dock, pit toilet, and hiking trail, along with the ruins of an 1890s homestead, but it is accessible only by boat.

In and beyond the SNAs, the park's cliff microhabitats host rare snails, crustaceans, ferns, and flowers. The forests are dominated by maple and beech. White cedars, many of them hundreds of years old, grow on the lakeside terraces and cling to the cliffs. At least 125 species of birds have been identified in the park, a popular destination for bird-watchers during spring migration time.

Also within the park's borders are nearly 20 miles of hiking trails, a 10-mile bike route, 12 miles of off-road biking trails, and extensive trail systems for cross-country skiing, snowshoeing, and snowmobiling. Some of the hiking trails are loops, but most are segments that can be connected for a large variety of hiking experiences. The two described in the Trail Guides provide a good sampling of the park's geological and natural features.

Geologists think that in this area, the Silurian sea must have been quiet and shallow, thriving with primitive life-forms that quietly lived and died to form the layers of rock that underlie the park. On your hikes and in the stone walls and walkways in the park, look for fossils, including those of honeycomb and chain corals, clamlike brachiopods, and the tracks of ancient worms and other crawling and burrowing creatures.

TRAIL GUIDE
Eagle Trail

This two-mile loop includes very difficult rocky, steep sections that require careful hiking for safety. Starting from the Eagle Panorama parking lot, the trail descends steeply down the bluff. After 0.2 mile, you begin to see dramatic rock outcroppings on the right. Getting closer to the lakeshore, the trail begins to level off but remains very rocky and slippery wherever the trail crosses wet areas fed by the many springs in the rocks above. The bluff here is heavily forested with cedar, hemlock, white pine, and birch, which along with the moss-covered rocks, give the trail an ancient feel.

After about 0.4 mile, the higher dolomite bluffs come into view on your right as you hike closer to the lakeshore cliffs and cobblestone beaches. At 0.7 mile you are walking right along the shore on an ancient terrace with the cliffs crowding closer. As you approach these cliffs, you see how the waves have worked them for centuries, eroding large angular chunks away from the cliff base. In places, the entire bluff seems to be sitting on dolomite pillars that act as archways into shallow caves (Figure 6.12).

Nearing the one-mile mark, you are hiking beneath the most dramatic palisades in the park. Look straight up and you see the concave underside of the uppermost cliffs, where chunks of rock will keep breaking away as the cliffs migrate inland inch by inch. At one point, you see Eagle Cave, the largest cave in the cliff, carved by the waves of Lake Algonquin around 10,000 years ago, now elevated 30 feet over the lakeshore (Figure 6.11).

6.12 These pillars of stone and the shallow caves behind them were carved by lake waves.

Beyond this breathtaking section of trail, at about 1.2 miles, you will find yourself in an unusually thick stand of cedar, hiking on the terrace next to Eagle Harbor. Shortly beyond that, the trail starts ascending away from the lake. Like the route down, the hike up the bluff is rocky and steep. At the 1.4-mile mark, the path tops the escarpment and levels off.

At 1.6 miles, you can take a stairway down to the Eagle Terrace Overlook, built on the site of an abandoned quarry, with a view of Eagle Harbor and Ephraim across the bay. The steps and broad stone terrace are made of slabs of dolomite, quarried, hauled, and put in place by the CCC workers. From here, the trail crosses Shore Road and meanders for 0.4 mile through the upland forest on a flat trail back to the Eagle Panorama parking lot.

TRAIL GUIDE

Hidden Bluff/Sunset/Skyline/Nicolet Bay Trails

This 4.5-mile hike combines segments of four trails to form a loop that starts and ends at the Nature Center. It is all easy hiking on well-groomed trails with a few gentle ups and downs. The trail junctions can be confusing, even for experienced hikers, so bring a map and be on the lookout for trail signage.

Start hiking west from the Nature Center on the Hidden Bluff Trail, which is actually a gravel bike trail but used by hikers as well. The trail parallels the feature for which it is named—a bluff that is being reclaimed by the forest. It is heavily eroded, so it does not have the sheer vertical faces that the park's palisades over the water have. It rises out of the woods like a ghost cliff that probably was much closer to the lakeshore thousands of years ago, before the lake level dropped and the land rebounded after the glacial retreat.

At the half-mile mark, the trail meets the Sunset Trail, another popular bike trail. This is also the junction with the Hemlock Trail, which takes a sharp left and ascends the bluff. The turn onto Sunset Trail is a right angle to the left, or south. Shortly after this junction, the cliffs to the left become more dramatic, rising 40 to 60 feet above the forest floor and more freshly eroded. You will pass through stands of beech, basswood, and maple with hemlock interspersed. As you gently ascend on this gravel bike trail, the white dolomite cliffs to your left become more dramatic, rising to 100 to 150 feet above the lake level (Figure 6.10). On your right, to the west, Green Bay comes into view. The trail is named for the gorgeous sunsets that many hikers and bikers have enjoyed on this stretch.

At about one mile from the start, the trail has made a gentle ascent to the bluff top from which you get a closer view of a dolomite outcropping and talus field below it. In another half mile, the trail passes through a thick stand of cedar and crosses Skyline Road at its junction with Middle Road. Stay on Sunset Trail, which after another 0.2 mile, crosses a leg of the Skyline Trail that goes back to the north. Pass this junction and keep going another 200 yards to the next junction with the east-west stretch of the Skyline Trail and turn left onto this trail, which is for hiking only.

After about a quarter mile, this trail crosses Middle Road and, after another tenth mile, enters a beautiful piney opening with a small stand of old hemlock surrounded by young pines and shrubs. The trail at this point is wide and flat with sandy soil underfoot covered deeply in pine needles. On the edge of this expansive clearing is a bench for resting—a good place to sit and enjoy this peaceful area of the park.

Shortly beyond this clearing, the trail begins ascending and becomes very rocky. The rounded nature of many of the rocks indicates that this might have been an old beach ridge. Here is where one of the ancient lakeshores sat as the icy waves of a glacial lake tumbled and washed the rocks for centuries. The trail soon levels off and comes to a Y-junction, with Nicolet Bay Trail going to the right and the Skyline Loop continuing to the left, having merged with the Hemlock Trail. The Nicolet Bay Trail is the one to take to complete the loop to the Nature Center. It is wide and flat with some roots and rocks—a mostly easy trail that descends gently to the north through a forest of extraordinarily large hemlock, basswood, and beech trees.

After about a third of a mile on the Nicolet Bay Trail, you cross Hemlock Road and, 200 yards farther on, meet a connector trail (labeled as such) coming from the east. The Nicolet Bay Trail veers left here, heading northwest, and shortly beyond, crosses Skyline Road. Across the road, you pick it up slightly to the left. As it approaches the Nature Center area, it skirts a very wet lowland, called a forest seep, where groundwater has filled a shallow basin. As a trail sign explains, this is an unusual forest ecosystem being preserved in the park. Shortly beyond this point, the trail crosses Hemlock Road again and then meets the Hidden Bluff Trail, on which you turn left to go back to the Nature Center and close the loop.

Newport State Park

Crossing from the Green Bay side of the Door Peninsula to the Lake Michigan side, you leave behind the cliffs soaring over Green Bay and follow the general downward slope of the land to the eastern shore where the cliffs are closer to the waters of Lake Michigan. While they are not strikingly high and dramatic, those cliffs possess a ragged beauty of their own (Figure 6.13) and in many places they merge with broad sandy beaches and forested sand dunes. These features are well displayed on the northeastern tip of the peninsula in Newport State Park.

This park, like Whitefish Dunes State Park, was developed at the site of an ancient bay on which sand dunes formed near the mouth to create a lesser bay

6.13 This ancient terrace in Newport State Park was created by the waves of Lake Michigan when the lake level was higher.

and a large lake just inland from Lake Michigan. This is the northernmost of a series of such formations that lie in valleys carved across the peninsula into 400-million-year-old Silurian dolomite.

The glaciers did their share of this valley erosion, starting close to 2 million years ago and covering the peninsula several times. While they left deep layers of till across most of eastern Wisconsin, only a thin layer lies over the peninsula. In the park area, the main evidence of glaciers is the scattering of glacial erratic stones and boulders. Farmers in the area used them to make stone fences long ago, and in the park, they have been used as border stones around parking areas. Look for basalt and granite boulders transported from the north, as well as smaller chunks of dolomite broken from the bedrock and moved across the peninsula by the Green Bay lobe of the glacier.

Other evidence of the glacier is the presence of ancient rocky shores of earlier versions of Lake Michigan. In the park, the most prominent of these are the beach ridges created by Lake Nipissing that lie inland from today's shoreline, 20 to 30 feet higher than the present lake level, now being reclaimed by the forest.

Europe Bay was originally an inlet that stretched beyond the shores of today's Europe Lake. Since the last glacier, Lake Michigan's level has been lower, as well as higher, than today's levels. By about 8,000 years ago, the lake had dropped at least 300 feet lower than the current level and Europe Bay likely was left high and dry. Within the next 1,500 years, as lake levels slowly rose, it had become a vast marsh, and by 6,000 years ago, it was again filled with water.

The replenished bay was a part of Lake Nipissing. On the north side of the bay was a rocky headland around which the waves and currents of the lake swirled. They moved massive amounts of sand and deposited it south of the headland, slowly forming a sand spit that eventually stretched southwest across the mouth of the bay as a semi-permanent sandbar. By 4,500 years ago, this sandbar became the site of growing sand dunes that closed off the inland body of water, creating Europe Lake. Over 2,000 years, the dunes grew to their present size as grasses and other pioneer plants, and later dune forests, took hold and stabilized the land. It is now a heavily forested isthmus separating Europe Lake and Europe Bay.

The park is named for the village of Newport, which once lay along the beach of Newport Bay straight east of Parking Area 3. In the late 1800s, it was a center for lumbering and shipping. Little remains of the village now, but starting with a boat pier built in 1881, it spread out to include several homes, a general store and post office, and a sawmill. During the next 40 years, the village produced tons of cordwood, railroad ties, posts, poles, and other products made from the trees being cut in the area. When the lumbering boom was over and the local forests were depleted, the village of Newport began to fade.

In 1919, German immigrant Ferdinand Hotz, who had become a successful Chicago merchant, purchased the land that is now Newport State Park. He was moved to do so by his disdain for the way the forests of Wisconsin had been cleared for timber during the previous decades. His goal was to protect the land and let nature restore it to the diverse forestland it had been before it was logged. Hotz was also a talented photographer, and his collection of photographs documenting the natural beauty and communities of Door County in the early 1900s has been preserved by the Wisconsin Historical Society.

The state bought the land from the Hotz estate in 1966 and in 1977 designated it a wilderness park with strict limits on development. For example, no campground was established, and the park instead has 17 secluded, primitive campsites accessible only by hiking trails. On the trails are signs that inform hikers about the park's geology, ecosystems, and flora and fauna.

In support of this wilderness approach, the Newport Wilderness Society formed in 1985, dedicated to promoting the study and preservation of the park's natural features and ecosystems, educating park visitors about these efforts, and supporting park managers. It is the counterpart to the friends groups that have organized around other state parks.

Newport State Park includes 2,400 acres of second-growth forests, wetlands, and meadows, 10 miles of shoreline on Lake Michigan, and 1 mile of shoreline on Europe Lake. The park also has more than 30 miles of trails, including 17 miles of off-road bicycling trails and 23 miles of cross-country skiing and snowshoeing trails, all of which are open for hiking.

TRAIL GUIDE
Lynd Point/Fern Loop Trail

This fascinating loop hike, about 2.5 miles long, combines two trails that are mostly level and easy to hike except for some rocks and roots on the trails, which are slippery in wet areas. Start by taking the Europe Bay Trail from the beach area near Parking Lot 3. This is the site of the vanished Village of Newport. A sign on the trail here informs you about the village.

Just over 0.3 mile from the parking lot is a junction with Lynd Point Trail, which goes to the right toward the lakeshore. It runs through stands of cedar, some of them hundreds of years old. All along the trails in this park, dolomite crops up, reminding you that the bones of this land are only inches under the soil.

Just past the half-mile mark you begin to get open views of the low cliffs over the shoreline, gradually being knocked apart by waves (Figure 6.13). If you proceed carefully, you can get a close look at this process. The flat terrace on which you will walk above the cliffs was created by Lake Michigan. Its waves have worked for centuries to clear this terrace of overlying glacial till and loose or softer rock. This part of the shore also has good exposures of a type of dolomite that contains numerous fossils. Look for signs of tubelike corals (*Syringopora*), honeycomb corals (*Favosites*) (Figure 6.14), and chain corals (*Halysites*) (Figure 6.15). An excellent pamphlet called *Geology of Newport State Park*, available at the park office, gives more information for fossil hunters.

At the one-mile mark, the trail passes campsites 1 and 2, and a quarter mile farther on, the trail rounds Lynd Point and heads northwest along the Europe Bay shoreline. You get more views of the rocky lakeshore all along this trail until the 1.75-mile point, where it turns inland and heads roughly southwest.

6.14 A fossil of a honeycomb coral (*Favosite*) found at Newport State Park. NEWPORT STATE PARK

6.15 A chain coral (*Halysite*) fossil found at Newport State Park. NEWPORT STATE PARK

6.16 An ancient shore of Lake Nipissing is now being reclaimed by the forest in Newport State Park.

The trail now crosses a damp hemlock lowland for a tenth of a mile to where you come to a moss-covered dolomite wall standing in the middle of the woods (Figure 6.16). This is an abandoned lakeshore—a broken, jagged wall 6 to 10 feet high—formed by Lake Nipissing. In the dim past, it was pounded for centuries by waves like those on today's shoreline 150 yards behind you. Now, this silent, abandoned lakeshore is being pulled apart not by relentless waves but by the steady, silent work of rain, frost, and vegetation clawing its way into every crack and crevice in the rock.

The trail runs along the base of the abandoned shore cliff for about a tenth of a mile to where it ascends the broken remnants of the wall, which at this point is disappearing into the forest floor. On a windy day, the steady rhythm of the rising and falling gusts of wind in the pines can sound like a ghostly echo of the waves that washed the ancient shore all those centuries ago.

At the two-mile mark, Lynd Point Trail ends at Europe Bay Trail, and the wide, flat Fern Trail continues across the junction, looping back toward the parking area. About a third of a mile from this junction, Fern Trail crosses a low wetland of cedar and ferns on several wide, well-built boardwalks. The helpful trail signs inform you that you are in a rare wetland microhabitat within a mixed hardwood forest. This and several other interpretive signs in the park are true to the former landowner Hotz's vision of a restored forest valued most for its natural systems and its beauty.

Rock Island State Park

Between the tip of the Door Peninsula and the Garden Peninsula, which juts southwest into the lake from Upper Michigan, is a northeast-trending series of islands extending like steppingstones between the two peninsulas. They form an archipelago that stretches across the mouth of Green Bay, and they figure prominently in the history of this part of the country. How, you might wonder, did they get there?

As with many questions about the lay of the land in northeastern Wisconsin, the answer takes us back to the formation of the Niagara Cuesta about 300 million years ago. Recall that the cuesta ran north across Wisconsin and curved around the Michigan Basin. To the west of the basin was land sloping off the Wisconsin Dome. The sloping rock layers were eroding to form the cuesta, with the escarpment—the ragged edge of harder eroding rock—on its west side and the gentler slope of uneroded rock on its east side.

Geologists think that, just before the glaciers arrived about 2 million years ago, rivers were draining northeast along both flanks of the cuesta, which at that time was a more continuous ridge between Wisconsin and Upper Michigan. Streams also formed in the southeast-trending cracks in the dolomite mantle over the cuesta, forming valleys there over millions of years. Because the west side of the cuesta was slightly higher than the east side, some streams on the west side broke through low points in the cuesta ridge to flow in these cross-cuesta valleys, draining southeast into the ancient Michigan Basin river. Such streams were thought to have flowed through low areas that now separate some of the islands in the archipelago northeast of Door County.

Over millions of years, these streams would have carved their valleys deeper. When the glaciers came, the succeeding masses of flowing ice sped up the erosion. Glacial lobes carved the basins deeper and during the last glaciation, the

Green Bay lobe finished the excavation of Green Bay, while the more massive Lake Michigan lobe carved the lake basin much deeper than the Green Bay basin. At the same time, the glaciers chipped away at the cuesta, making its stream valley gaps larger. On a detailed map of the archipelago, you can see that a few of its islands have a similar shape, something like a football oriented northwest and southeast. It seems likely that a glacier flowing southeast, as that margin of the Green Bay lobe did, would carve the islands in this way.

When the glacier stopped advancing and began to melt, it had created what is called a hanging valley in the Green Bay basin. It happens when a major valley is eroded deeper than its tributary valley. The Green Bay lobe carved Green Bay to depths of 100 to 144 feet while the Lake Michigan lobe plowed the lake bottom to depths of 500 feet and more.[9] Imagine the torrential flows of meltwater from the higher to the lower basin when the ice masses softened and melting accelerated. Ice water and icebergs would have gushed through the gaps in the cuesta, tearing away more rock and widening those gaps.

As the glacial lakes drained away and lake levels dropped, the remaining high points along the ravaged cuesta emerged to become islands in the postglacial lake. Over thousands of years during the glacial retreat, glacial lakes dropped thick layers of sand and other eroded sediments around these islands. If you could now drain away all the water, scoop out these deep layers of sediment, and rise to a bird's-eye view of the area, you would see a thin, ragged remainder of the Niagara Escarpment stretching from the Door Peninsula to the Garden Peninsula with a few stubborn high peaks scattered along its length.

But put the sediment back and fill up the lake, for we will now zoom down to lake level and take a closer look at one of those high peaks, today called Rock Island (Figure 6.17). It lies just north of Washington Island as part of a cluster of islands on the Wisconsin side of the state line that also includes Plum and Detroit Islands and several smaller islets. To the north of Rock Island on the Michigan side of the line lie St. Martin, Poverty, and Summer Islands, and several smaller islands stretch toward the Garden Peninsula.

As with all the islands, Rock Island is part of the cuesta, so much of its western side is a sheer cliff towering as high as 140 feet over the lake at the northwest corner. Its interior rises to 200 feet and then slopes gently down to the east side where the cliffs are no higher than 60 feet over the water. At the southeast corner, the cliffs dwindle to merge with a cobblestone beach where much of the earliest settlement of the island took place. Across the island on the southwest

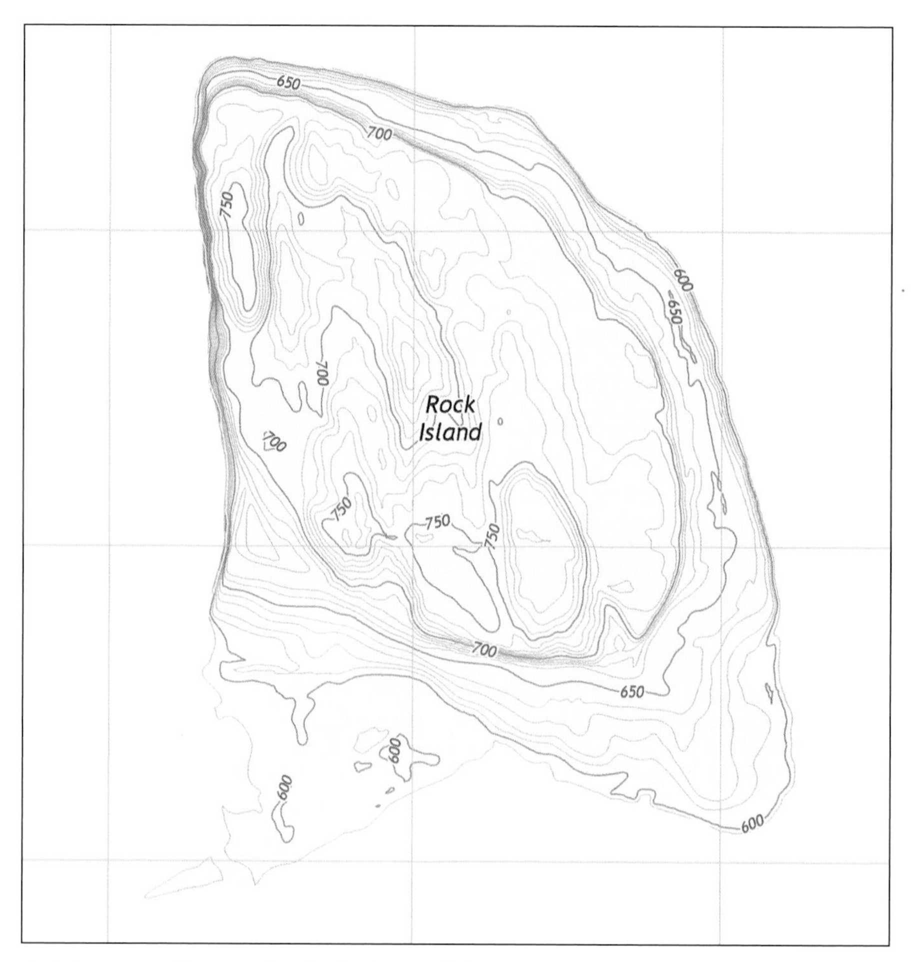

6.17 A topographic map of Rock Island, part of the Niagara Cuesta. Note that the western shore is higher than the eastern shore due to the slope of the cuesta. US GEOLOGICAL SURVEY

corner, waves and currents have moved sand for thousands of years to build a spit extending like a tail into the lake where visitors now enjoy a sandy beach area. Scattered across the island's interior are erratic boulders of granite and basalt dropped by the glaciers, along with a very thin layer of till on which the forest we see today took hold.

Archaeologists have found evidence of Native Americans living on the island at least 2,000 years ago and of repeated visits by peoples of the Middle and Late Woodland tradition—hunters and gatherers who probably made fishing camps on the island during warmer seasons. More modern tribes that inhabited the island were Huron, Petun, and Odawa, who were primarily traveling through and did not stay long. The Potawatomi had established a village occupied periodically between 1670 and 1730. When the first white settlers arrived in the 1830s, an Ojibwe village was located on the east shore, actively fishing the waters.[10]

When early European explorers and fur traders were making the Grand Traverse on Lake Michigan, Rock Island could have served as a stopping place. Some historians speculate that part of their route could have included the passage north of Rock Island, which would have allowed the travelers to use the southwest end of the island as a rest stop.

European settlement on Rock Island began in the early 1840s with the building of a pier on the southwest corner of the island. Later, a fishing village was built on the southeast corner—possibly the first European settlement on all of the Door Peninsula.[11] In fact, a European immigrant community and economy developed on Rock Island long before Washington Island saw such activity. The population of Rock Island was recorded at 218 in 1855. Now only three small cemeteries commemorate the once-active rural community.

Throughout the archipelago are numerous shallows and shoals that make passage among the islands hazardous for ships of any size. The strait between the Door Peninsula and Washington Island is called Porte des Morts, French for Death's Door, for which the Door Peninsula was also named. Porte des Morts and Rock Island Passage northeast of Rock Island both hold narrow zones of safe passage, which became the scene of many shipwrecks on the shoals of the archipelago before the US government built lighthouses on the islands. The first of these, and the first lighthouse in Wisconsin, was built in 1836 on Rock Island's northwest bluff, called Pottawatomie Point (a misspelling of the tribal name).

Life on Rock Island was tough, and as Washington Island and points south began to attract more development, the residents of Rock Island moved and the population shrank. The last family left the island around 1890.

In 1910, Chicagoan Chester Thordarson, a brilliant and successful inventor and manufacturer of electrical transmission equipment, bought up all but a few tracts of land on the island. He was an Icelandic immigrant, and the island reminded him of his home country. He first restored a house in the abandoned east shore village area and settled there, and then built a water tower with the intension of restoring the village. However, the water tower—still as solid as it ever was—was never used because Thordarson decided to develop the southwest corner of the island instead.

The most prominent of the several structures Thordarson built is the iconic Viking Hall (Figure 6.18), a massive meeting hall with a boathouse beneath it. It was built of dolomite blocks and glacial erratic boulders as a display of Icelandic architecture. Thordarson used the hall to entertain his family, friends, and employees during vacation retreats on the island. It housed ornate, hand-carved tables, chairs, and benches and scores of bookshelves on which he stored his impressive collection of rare books. Thordarson was a Renaissance man, reading voraciously about physics, chemistry, zoology, agriculture, cooking, medicine, and many other topics. When he died in 1945, his prized book collection went to the University of Wisconsin. Except for the books, the hall remains much as it was, now preserved as a museum.

Other structures he built included his son's house (now a state park ranger office), a greenhouse (now a park shelter), a cookhouse, and an elaborate cedar gate on top of the hill to the north overlooking it all, now partly decayed but fenced off and preserved. Thordarson was also a proud steward of the land. He studiously avoided disturbing all but the 30 acres he developed on one corner of the island, leaving the rest to nature. For this, he received recognition as an outstanding conservation leader.

6.18 Viking Hall, used by its builder Chester Thordarson as a gathering place, library, and boathouse.

6.19 A section of a wooded trail in the peaceful and minimally developed Rock Island State Park.

When Thordarson died in 1945, a number of visionaries had plans for developing the island. Some hoped to parcel it into luxury estates and add a marina for the new owners' yachts. One plan envisioned a flow of tourists traveling on a causeway to be built between Washington and Rock Islands. Thordarson's heirs rejected all such plans and, in 1964, sold the land to the state of Wisconsin for use a state park.[12]

Rock Island State Park now occupies 912 acres, all but a quarter acre of the island. The latter is the site of a solar-powered navigation light owned by the US Coast Guard right next to the Pottawatomie Lighthouse on the northwest bluff. The original 1836 lighthouse crumbled and was replaced in 1858. While it no longer lights the channel, it is now a museum available for touring in spring, summer, and fall. Viking Hall has also been maintained well and houses an excellent, informative display of the findings of archaeologists about the earliest human inhabitants of the island. These attractions are all maintained and operated by Friends of Rock Island, a volunteer group whose mission is to support the Department of Natural Resources in maintaining the state park.

Rock Island is a wilderness park accessible by a passenger-only ferry that runs seasonally from Washington Island (which is served by car ferries running year-round from the Northport Pier in Door County). Private boats are also allowed to use the dock and boathouse. The island has no roads, and no cars or bikes are allowed, the only vehicle being an ATV used by park rangers. The park has 40 primitive campsites, all situated close to the lakeshore on the southeast and southwest corners of the island, with a pit toilet located near each campsite cluster. About 6 miles of shoreline and 10 miles of trails are available for hiking.

This state park has done well to preserve both Thordarson's estate and the surrounding wild land that he loved and protected. It does not afford stunning views of spectacular cliffs or waterfalls, but its trails lie easily in a quiet, undisturbed forest. It affords a peaceful, relaxing time in an ancient place and reminds you that such simple beauty is also quite spectacular (Figure 6.19).

TRAIL GUIDE
Thordarson Trail

Rock Island has three main hiking trails, two that transect the park and one that lies along the perimeter. The latter affords views of the lake and the chance to see as much of the island as possible. It is a generally wide, level, and easy trail with the exception of a few steep stretches.

The tongue-in-cheek name Michigan Avenue was given to the straight, wide, grassy path from the dock and boathouse to the camping area. Beginning from its junction with Thordarson Trail, hike east along the south shore of the island. After a third of a mile, you will arrive at a beach area—a wide clearing stretching down to the lake, marked on the state park map as Rutabaga Field. It may be one of the sites where a family tried to eke out a living by farming in the late 1800s.

Within a half mile, the trail passes a low ridge on the left, likely the site of an ancient beach built by the waves of Lake Nipissing. In another tenth of a mile, the trail passes by an amazingly thick shoreline cedar forest in which you can barely see daylight between the tree trunks in some places. This is where a trail to two primitive campsites cuts to the right, toward the lakeshore, while the main trail angles away from the lake and goes to higher ground.

At 1.25 miles, the trail passes a site where a fishing village stood more than 150 years ago. All that remains of it are piles of stones and depressions where houses and barns once stood and where wells were dug. A trail sign has more information about the village. A little beyond this site, the trail ascends out of

6.20 The Pottawatomie Lighthouse, now a museum, was built in 1858 to replace the original built in 1836—Wisconsin's first lighthouse.

the beach area and passes the water tower with fireplace that Thordarson built. It never actually served as a water tower but now provides a shady resting place. Here the trail narrows from road width to a foot trail.

Shortly beyond the tower, the trail crosses a small area of dune forest on deep sand and rises slightly to run along the low east-side bluff over the lake. Here bedrock lies close to the surface and crops up as slabs of dolomite all along the trail. The bluff on this side of the island reaches a height of about 60 feet.

At the 2.4-mile mark, the trail passes a cemetery in the woods to the left. It holds the remains of hardy pioneers who fought to make a living on this rocky island. At this point, the trail has started angling northwest through woods, not close to the lakeshore, but parallel to the northeast side of the island. After another third of a mile, it rises moderately, becoming more rocky and rugged. At the 2.8-mile mark is a dolomite outcropping on the right that was probably scraped smooth by glacial ice thousands of years ago. Look for such polished rock and erratic boulders in this, one of the higher areas of the park. A little farther along on this trail is a resting bench with a view to the northeast.

At the three-mile mark, you come to a junction with the Fernwood Trail, one of the two trails that cross the island, named for the picturesque fern communities through which it passes. A half mile farther on is another cemetery and, just beyond that, the lighthouse and picnic area. It is worthwhile to take a tour of the lighthouse (Figure 6.20), hosted by docents who live there for part of the season. You can get a good sense of what life was like for the lighthouse keeper and for the little community on Rock Island during the 1800s. You are allowed to climb stairs to the top of the tower to see the workings of the now-inoperative light and the panoramic view to the north that takes in other islands of the archipelago.

From the lighthouse, you make a moderately steep climb to one of the high points on the island where the trail levels off and heads south along the west side of the island. After descending gently, the trail arrives at the Gate, a curious-looking structure built out of cedar tree trunks and stumps. It was part of the deer fence that Thordarson erected around the developed southwest corner of the island, intending to keep deer out of the area and to keep them wild. The Gate is reminiscent of Viking architecture and intended to be highly visible from below, serving as a symbol of Thordarson's pride and love for his island. From here, you descend to the large, grassy area that contained Thordarson's estate. At the 5-mile mark, you will pass Viking Hall on the last stretch of the trail.

Acknowledgments

An important resource for me in writing this book was the information and services provided by the various Friends groups. These volunteer groups formed around many of the parks for the purposes of helping the Department of Natural Resources (DNR) in protecting and improving the parks, teaching people about their importance, and helping people to use them sustainably. To that end, volunteers play many roles, including campground hosts, nature center hosts, educators, trail workers, carpenters, researchers, planners, and fund-raisers. We who enjoy the state parks owe them a huge debt of gratitude.

It is hard to know where to draw the line in naming all of the people who helped me through the long, exciting, and sometimes trying process of writing this book. I first want to thank Kathy Borkowski and Kate Thompson for putting their faith in this project and allowing me the honor of working with the Wisconsin Historical Society Press. At the press, developmental editor Erika Wittekind gave invaluable advice and made great improvements to the manuscript. In my travels, I got information and guidance from many friendly, hardworking state park employees, DNR employees, members of Friends groups, and Ice Age Trail volunteers, notably Erin Brown, William Bursaw, John Carrier, Kent Goeckermann, Becky Green, Lois Hanson, Michelle Hefty, John Helling, Susan Johansen, Kevin Keeffe, Jackie Scharfenberg, Dan Schuller, Colleen Tolliver, and Heather Wolf. Equally helpful was the guidance of professors Bob Dott of the University of Wisconsin–Madison, Tom Fitz of Northland College in Ashland, and Keith Montgomery of the University of Wisconsin–Marathon County.

In the daily research and writing grind, I got invaluable advice from Jerry Apps, whose books, teachings, sense of humor, and boundless energy serve as true inspiration. I owe a large debt of gratitude to my long-time friend and mentor Tyler Miller, who developed the textbook that has long served as the model for most introductory college environmental science courses. It has been my great privilege to work with him as a coauthor on later editions of that book.

To my friends and family who cheered me on, put me up for a night, gave me a free meal or two, or offered advice on what to write about, I can't thank you enough. Thanks also go to my son, Will, for his help with photos and to my daughter, Katie, for her thought-provoking questions. Last, and certainly most, from the bottom of my heart, my love and thanks go to Gail, my patient friend, wife, and dedicated fellow tree-hugger who has never wavered in her enthusiasm and generous support for this project.

Notes

Chapter 1

1. Different authors use varying time frames for the eras and periods of Earth's ancient history. I rely on a blend of those presented by geologists Robert H. Dott Jr., John W. Attig, and David M. Mickelson.
2. Gene L. LaBerge, *Geology of the Lake Superior Region* (Tucson: Geoscience Press, 2004), 61.
3. Robert H. Dott Jr. and John W. Attig, *Roadside Geology of Wisconsin* (Missoula, MT: Mountain Press Publishing, 2004), 41–44; and J. A. Luczaj, "Geology of the Niagara Escarpment in Wisconsin," *Geoscience Wisconsin* 22, no. 1 (2013): 6.
4. Dott and Attig, *Roadside Geology of Wisconsin*, 48–50.
5. LaBerge, *Geology of the Lake Superior Region*, 149.
6. Dott and Attig, *Roadside Geology of Wisconsin*, 12; and Gwen M. Schulz, *Wisconsin's Foundations* (Madison: University of Wisconsin Press, 2004), 10.
7. Dott and Attig, *Roadside Geology of Wisconsin*, 20; and LaBerge, *Geology of the Lake Superior Region*, 225.
8. David M. Mickelson, Louis J. Maher Jr., and Susan L. Simpson, *Geology of the Ice Age National Scenic Trail* (Madison: University of Wisconsin Press, 2011), 17–18.
9. LaBerge, *Geology of the Lake Superior Region*, 243.
10. Dott and Attig, *Roadside Geology of Wisconsin*, 22–23; and LaBerge, *Geology of the Lake Superior Region*, 244.
11. Donald M. Waller and Thomas P. Rooney, eds., *The Vanishing Present: Wisconsin's Changing Lands, Water, and Wildlife* (Chicago: University of Chicago Press, 2008), 65–66.
12. LaBerge, *Geology of the Lake Superior Region*, 242-243 and 293–294.

Chapter 2

1. Gwen M. Schulz, *Wisconsin's Foundations* (Madison: University of Wisconsin Press, 2004), 40–41.
2. Some state park literature says the Douglas Fault occurred 500 million years ago. The exact time is not known and geologists differ on this point, but I use Dott and Attig's interpretation that the faulting began around 900 million years ago. See Robert H. Dott Jr. and John W. Attig, *Roadside Geology of Wisconsin* (Missoula, MT: Mountain Press Publishing, 2004), 83–84.

3. For the full story of the CCC's work in Pattison State Park, see the booklet *The Story of Camp Pattison* by John V. Semo (Wisconsin DNR, 2003), which is available at the park office.
4. Paul C. Tychsen, *Geology of Amnicon Falls State Park Field Guide* (Superior: University of Wisconsin–Superior, Department of Geology, 1975), 15.
5. This trail description is based on my hikes in the park, but I am supplementing it with information from the helpful, illustrated *Amnicon Falls Geology Walk* pamphlet, published by the state Department of Natural Resources and available at the park office.
6. The *Amnicon Falls Geology Walk* pamphlet gives more details on this feature of the park (Stop 4).
7. Robert H. Dott Jr. and John W. Attig, *Roadside Geology of Wisconsin* (Missoula, MT: Mountain Press Publishing, 2004), 115.
8. G. T. Owen, *The Geology of Copper Falls State Park* (Madison: Wisconsin Conservation Department, 1938), 1.
9. W. F. Cannon, S. W. Nicholson, R. E. Zartman, Z. E. Peterman, and D. W. Davis, "Kallander Creek Volcanics—A Remnant of a Keweenawan Central Volcano Centered Near Mellen, Wisconsin," *Proceedings of the 39th Annual Meeting of the Institute on Lake Superior Geology* 39, no. 1 (1993): 20–21.
10. Tom Fitz, telephone interview with the author, Feb. 22, 2017.
11. Kent Goeckermann, email correspondence with the author, Feb. 24, 2017.
12. Owen, *The Geology of Copper Falls State Park*, 4.
13. Allison Mills, Drew Cramer, and Tom Fitz, "Field Trip 6: Geology Hike at Copper Falls State Park," *Proceedings of the 57th Annual Meeting of the Institute on Lake Superior Geology*, 57 (2010): 7.
14. David M. Mickelson, Louis J. Maher Jr., and Susan L. Simpson, *Geology of the Ice Age National Scenic Trail* (Madison: University of Wisconsin Press, 2011), 68.
15. Dan Schuller, telephone interview with the author, April 4, 2017.
16. Dott and Attig, *Roadside Geology of Wisconsin*, 59.
17. Alonzo W. Pond, *Interstate Park and Dalles of the St. Croix* (St. Croix Falls, WI: The Standard-Press, 1937), 21.
18. Dott and Attig, *Roadside Geology of Wisconsin*, 93–95.

Chapter 3

1. Gene L. LaBerge, *Geology of the Lake Superior Region* (Tucson: Geoscience Press, 2004), 242–243.
2. Eric C. Carson and J. Elmo Rawling III, *Late Cenozoic Evolution of the Lower Wisconsin River Valley* (Madison: University of Wisconsin–Extension, Wisconsin Geological and Natural History Survey, 2015), 3–5.
3. John T. Curtis, *The Vegetation of Wisconsin: An Ordination of Plant Communities* (Madison: University of Wisconsin Press, 1959), 14.
4. David Mickelson, "A Hike into the Past: Five Days in the Ice Age," in Bart Smith, *Along Wisconsin's Ice Age Trail* (Madison: University of Wisconsin Press, 2008), 11.
5. Patty Loew, *Indian Nations of Wisconsin*, 2nd ed. (Madison: Wisconsin Historical Society Press, 2014), 45–46.
6. "The Indigenous Women Miners of the Driftless Area," *An Indiginous History of North America*, Feb. 18, 2015, indigenous history.wordpress.com/2015/02/18/the -indigenous-women-miners-of-the -driftless-area.
7. John Carrier, interview with the author, Sept. 30, 2015, Perrot State Park, Trempealeau, WI.

8. LaBerge, *Geology of the Lake Superior Region*, 269–270.
9. Lawrence Martin, *The Physical Geography of Wisconsin* (Madison: University of Wisconsin Press, 1965), 66.
10. Robert H. Dott Jr. and John W. Attig, *Roadside Geology of Wisconsin* (Missoula, MT: Mountain Press Publishing, 2004), 161–162.

Chapter 4

1. Gene L. LaBerge, *Geology of the Lake Superior Region* (Tucson: Geoscience Press, 2004), 117–118.
2. Robert H. Dott Jr. and John W. Attig, *Roadside Geology of Wisconsin* (Missoula, MT: Mountain Press Publishing, 2004), 196.
3. Ibid., 203–205.
4. *The Land with Jerry Apps* (Wisconsin Public Television, 2015); and Jerry Apps, personal communication with author, Madison, Jan. 14, 2017.
5. Dott and Attig, *Roadside Geology of Wisconsin*, 194.
6. Gwen M. Schulz, *Wisconsin's Foundations* (Madison: University of Wisconsin Press, 2004), 160.
7. Lawrence Martin, *The Physical Geography of Wisconsin* (Madison: University of Wisconsin Press, 1965), 59.
8. Lange, Kenneth I., *Song of Place: A Natural History of the Baraboo Hills* (Baraboo, WI: Ballindalloch Press, 2014), 516.
9. The whole story of the CCC at Devil's Lake is told eloquently by Robert J. Moore in his book *Devil's Lake, Wisconsin, and the Civilian Conservation Corps* (Charleston: The History Press, 2011).
10. Lange, *Song of Place*, 590.
11. Dott and Attig, *Roadside Geology of Wisconsin*, 216.
12. In their book *Roadside Geology of Wisconsin*, pp. 222–223, Dott and Attig give a superb account of the ancient and more recent history of the glen.
13. Martin, *The Physical Geography of Wisconsin*, 349.
14. For an excellent, detailed description of how erosion picked apart the sandstone mantle, leaving the mesas, buttes, and pinnacles of the Driftless Area, see Schultz, *Wisconsin's Foundations*, 99–114.
15. David M. Mickelson, Louis J. Maher Jr., and Susan L. Simpson, *Geology of the Ice Age National Scenic Trail* (Madison: University of Wisconsin Press, 2011), 231.
16. Dott and Attig, *Roadside Geology of Wisconsin*, 231.
17. Heather Wolf, telephone interview, April 26, 2017.

Chapter 5

1. For more details on these vanished mountain ranges, see Gwen M. Schulz, *Wisconsin's Foundations* (Madison: University of Wisconsin Press, 2004), 60–61.
2. Lawrence Martin, *The Physical Geography of Wisconsin* (Madison: University of Wisconsin Press, 1965), 257.
3. David M. Mickelson, Louis J. Maher Jr., and Susan L. Simpson, *Geology of the Ice Age National Scenic Trail* (Madison: University of Wisconsin Press, 2011), 61.
4. Ibid., 68.
5. Ibid., 46.
6. Schulz, *Wisconsin's Foundations*, 159.
7. Mickelson, Maher, and Simpson, *Geology of the Ice Age National Scenic Trail*, 72.
8. Schulz, *Wisconsin's Foundations*, 145.
9. For more information on the trail, the work of the Ice Age Trail Alliance, and options for joining and volunteering, see www.iceagetrail.org.

10. Robert A. Birmingham and Lynne G. Goldstein, *Aztalan: Mysteries of an Ancient Indian Town* (Madison: Wisconsin Historical Society Press, 2005), 1.
11. Ibid.
12. Patty Loew, *Indian Nations of Wisconsin*, 2nd ed. (Madison: Wisconsin Historical Society Press, 2014), 7.
13. Birmingham and Goldstein, *Aztalan*, 53.
14. For more detail on the town layout, the social structure of the community, and daily life for the residents of Aztalan, see Birmingham and Goldstein, *Aztalan*.
15. Mickelson, Maher, and Simpson, *Geology of the Ice Age National Scenic Trail*, 114–116.
16. Ibid., 117.
17. Ibid., 121–125.
18. Ibid., 76.
19. Robert H. Dott Jr. and John W. Attig, *Roadside Geology of Wisconsin* (Missoula, MT: Mountain Press Publishing, 2004), 304.
20. Retired High Cliff State Park naturalist Cynthia Mueller provided a complete, detailed, and colorful description of the vanished town of High Cliff in "The Lost Town—High Cliff, Wisconsin," in *Impact Magazine*, published in 2002 by the Wisconsin Parks and Recreation Association, reprints of which are now available at the park office.
21. Loew, *Indian Nations of Wisconsin*, 47–48.

Chapter 6

1. Gene L. LaBerge, *Geology of the Lake Superior Region* (Tucson: Geoscience Press, 2004), 135–138.
2. Robert H. Dott Jr. and John W. Attig, *Roadside Geology of Wisconsin* (Missoula, MT: Mountain Press Publishing, 2004), 65.
3. LaBerge, *Geology of the Lake Superior Region*, 272.
4. David M. Mickelson, Louis J. Maher Jr., and Susan L. Simpson, *Geology of the Ice Age National Scenic Trail* (Madison: University of Wisconsin Press, 2011), 88.
5. William H. Tishler, *Door County's Emerald Treasure: A History of Peninsula State Park* (Madison: University of Wisconsin Press, 2006), 141–142.
6. Ibid., 141.
7. *Potawatomi State Park Visitor Guide* (Madison: Wisconsin Department of Natural Resources, 2016), 7.
8. For a colorful and complex history of the park, see Tishler, *Door County's Emerald Treasure*.
9. Lawrence Martin, *The Physical Geography of Wisconsin* (Madison: University of Wisconsin Press, 1965), 239–241.
10. Conan Bryant Eaton, *Rock Island* (Washington Island, WI: Jackson Harbor Press, 2002), 31.
11. Ibid., 31–34.
12. For a compelling and more complete history of Rock Island, see Eaton, *Rock Island*.

Selected Bibliography

Aldrich, Henry R. "The Geology of the Gogebic Iron Range of Wisconsin." (Madison: Wisconsin Geological and Natural History Survey, 1929).

Attig, John W., Lee Clayton, Kenneth I. Lange, and Louis J. Maher. *The Ice Age Geology of Devil's Lake State Park*. Madison: Wisconsin Geological and Natural History Survey, 1990.

Bailey, Bill. *Wisconsin State Parks, State Forests, and Recreation Areas*. Freeland, MI: Glovebox Guidebooks of America, 2003.

Birmingham, Robert A., and Lynne G. Goldstein. *Aztalan: Mysteries of an Ancient Indian Town*. Madison: Wisconsin Historical Society Press, 2005.

Busch, Jane Celia. *People and Places: A Human History of the Apostle Islands*. Omaha, NE: Midwest Regional Office, National Park Service, 2005.

Cannon, W. F., S. W. Nicholson, R. E. Zartman, Z. E. Peterman, and D. W. Davis. "Kallander Creek Volcanics—A Remnant of a Keweenawan Central Volcano Centered Near Mellen, Wisconsin." *Proceedings of the 39th Annual Meeting of the Institute on Lake Superior Geology* 39, no. 1 (1993): 20–21.

Carson, Eric C., and J. Elmo Rawling III. *Late Cenozoic Evolution of the Lower Wisconsin River Valley*. Madison: Wisconsin Geological and Natural History Survey, 2015.

Clayton, Lee. *Pleistocene Geology of the Superior Region, Wisconsin*. Madison: University of Wisconsin–Extension, Wisconsin Geological and Natural History Survey, 1984).

Clayton, Lee, and John W. Attig. *Glacial Lake Wisconsin*. Boulder, CO: Geological Society of America, 1989.

Clayton, L., and S. R. Moran. "Chronology of Late Wisconsinan Glaciation in Middle North America." *Quaternary Science Reviews* 1, no. 1 (1982): 55–82.

Curtis, John T. *The Vegetation of Wisconsin: An Ordination of Plant Communities*. Madison: University of Wisconsin Press, 1959.

Dixon, Dougal. *The Practical Geologist*. Edited by Raymond L. Bernor. New York: Simon & Schuster/Fireside, 1992.

Dott, Robert H., Jr., and John W. Attig. *Roadside Geology of Wisconsin*. Missoula, MT: Mountain Press Publishing, 2004.

Dott, Robert H., Jr., and Roger L. Batten. *Evolution of the Earth*, 4th ed. New York: McGraw-Hill, 1988.

Eaton, Conan Bryant. *Rock Island*. Washington Island, WI: Jackson Harbor Press, 2002.

Hoffman, Randy. *Wisconsin's Natural Communities*. Madison: University of Wisconsin Press, 2002.

Ice Age Trail Guidebook (Cross Plains, WI: Ice Age Trail Alliance, 2014).

LaBerge, Gene L. *Geology of the Lake Superior Region*. Tucson: Geoscience Press, 2004.

Lange, Kenneth I. *Song of Place: A Natural History of the Baraboo Hills*. Baraboo, WI: Ballindalloch Press, 2014.

Loew, Patty. *Indian Nations of Wisconsin*, 2nd ed. Madison: Wisconsin Historical Society Press, 2014.

Luczaj, J. A. "Geology of the Niagara Escarpment in Wisconsin." *Geoscience Wisconsin* 22, no. 1 (2013).

Martin, Lawrence. *The Physical Geography of Wisconsin*. Madison: University of Wisconsin Press, 1965.

McGraw-Hill Dictionary of Geology & Mineralogy, 2nd ed. New York: McGraw-Hill, 2003.

Mickelson, David M., Louis J. Maher Jr., and Susan L. Simpson. *Geology of the Ice Age National Scenic Trail*. Madison: University of Wisconsin Press, 2011.

Mills, Allison, Drew Cramer, and Tom Fitz. "Field Trip 6: Geology Hike at Copper Falls State Park." *Proceedings of the 57th Annual Meeting of the Institute on Lake Superior Geology* 57, no. 2 (2010): 97–110.

Moore, Robert J. *Devil's Lake, Wisconsin, and the Civilian Conservation Corps*. Charleston: The History Press, 2011.

Montgomery, Keith. "A Geologic History of Rib Mountain, Wisconsin." University of Wisconsin Colleges, June 12, 2017, http://pages.uwc.edu/keith.montgomery/ribmtn/ribmtn.htm.

Owen, G. T. *The Geology of Copper Falls State Park*. Madison: Wisconsin Conservation Department, 1938.

Paull, Rachel K., and Richard A. Paull. *Wisconsin and Upper Michigan*. Dubuque, IA: Kendall/Hunt Publishing, 1980.

Pielou, E. C. *After the Ice Age: The Return of Life to Glaciated North America*. Chicago: University of Chicago Press, 1991.

Pond, Alonzo W. *Interstate Park and Dalles of the St. Croix*. St. Croix Falls, WI: The Standard-Press, 1937.

Purinton, Richard. *Thordarson and Rock Island*. Washington Island, WI: Island Bayou Press, 2013.

Schultz, Gwen M. *Wisconsin's Foundations*. Madison: University of Wisconsin Press, 2004.

Schulz, Klaus J., and William F. Cannon. *The Penokean Orogeny in the Lake Superior Region*. Reston, VA: US Geological Survey, 2007.

Smith, Bart. *Along Wisconsin's Ice Age Trail*. Madison: University of Wisconsin Press, 2008.

Tishler, William H. *Door County's Emerald Treasure: A History of Peninsula State Park*. Madison: University of Wisconsin Press, 2006.

Trewartha, Glenn T. *Elements of Geography*. New York: McGraw Hill, 1967.

Tychsen, Paul C. *Geology of Amnicon Falls State Park Field Guide*. Superior: University of Wisconsin–Superior, Department of Geology, 1975.

Van Hise, Charles Richard, and Charles Kenneth Leith. *Geology of the Lake Superior Region*. Washington, DC: US Geological Survey, 1911.

Waller, Donald M., and Thomas P. Rooney, eds. *The Vanishing Present: Wisconsin's Changing Lands, Water, and Wildlife*. Chicago: University of Chicago Press, 2008.

Index

Page numbers in *italics* refer to illustrations.

About the Author

Scott Spoolman is a science writer who has focused on the environmental sciences, especially those stories of science and the environment related to Wisconsin and surrounding states. After earning a master's degree from the University of Minnesota School of Journalism, he worked for several years as an editor in the publishing industry, specializing in textbooks and other educational materials. Since 1996, he has worked as a freelance writer and editor for a variety of outlets and has coauthored several editions of a series of environmental science textbooks. Throughout his life he has enjoyed exploring the forests and waters of the upper Midwest, as well as the mountains of the western and eastern United States. His passion for the outdoors led him to develop an avid interest in geology and natural history, especially in his home state of Wisconsin.